C0-AWO-201

Plant-animal interactions in Mediterranean-type ecosystems

Tasks for vegetation science 31

The titles published in this series are listed at the end of this volume.

Plant-animal interactions in Mediterranean-type ecosystems

Edited by

M. ARIANOUTSOU and R.H. GROVES

Kluwer Academic Publishers
Dordrecht / Boston / London

Library of Congress Cataloging-in-Publication Data

Plant-animal interactions in Mediterranean-type ecosystems / editors,
Margarita Arianoutsou, R.H. Groves.
p. cm. -- (Tasks for vegetation science ; v. 31)
Papers outgrowth of a conference held at Maleme, Crete, Greece,
Sept. 23-27, 1991.
Includes index.
ISBN 0-7923-2470-6 (HB : acid free paper)
1. Animal-plant relationships--Congresses. 2. Mediterranean
climate--Congresses. 3. Biotic communities--Congresses.
I. Arianoutsou, Margarita. II. Groves, R. H. III. Series: Tasks
for vegetation science ; 31.
QH549.5.P54 1994
574.5'247--dc20 93-6345

ISBN 0-7923-2470-6

Published by Kluwer Academic Publishers,
P.O. Box 17, 3300 AA Dordrecht, The Netherlands.

Kluwer Academic Publishers incorporates the publishing programmes
of D. Reidel. Martinus Nijhoff, Dr W. Junk and MTP Press.

Sold and distributed in the U.S.A. and Canada
by Kluwer Academic Publishers,
101 Philip Drive, Norwell, MA 02061, U.S.A.

In all other countries, sold and distributed
by Kluwer Academic Publishers Group,
P.O. Box 322, 3300 AH Dordrecht, The Netherlands.

Printed on acid-free paper

Printed in the Netherlands

Table of contents

Preface *by* F. di Castri vii

Introduction *by* M. Arianoutsou and R.H. Groves ix

Historical introduction

Chapter 1. C.A. Thanos, Aristotle and Theophrastus on plant–animal interactions 3

Community structure

Chapter 2. R.L. Specht, Species richness of vascular plants and vertebrates in relation to canopy productivity 15

Chapter 3. J. Blondel and P.C. Dias, Summergreenness, evergreenness and life history variation in Mediterranean Blue Tits 25

Chapter 4. A. Legakis, Community structure and species richness in the Mediterranean-type soil fauna 37

Chapter 5. M.L. Cody, Bird diversity within and among Australian heathlands 47

Chapter 6. D.T. Bell, Plant community structure in southwestern Australia and aspects of herbivory, seed dispersal and pollination 63

Triangular relationships

Chapter 7. H.A. Mooney and R.J. Hobbs, Resource webs in Mediterranean-type climates 73

Chapter 8. B.B. Lamont, Triangular trophic relationships in Mediterranean-climate Western Australia 83

Herbivory

Chapter 9. N.G. Seligman and A. Perevolotsky, Has intensive grazing by domestic livestock degraded Mediterranean Basin rangelands? 93

Chapter 10. P.W. Rundel, M.R. Sharifi and A. Gonzalez-Coloma, Resource availability and herbivory in *Larrea tridentata* 105

Chapter 11. R. Ginocchio and G. Montenegro, Effects of insect herbivory on plant architecture 115

Pollination

Chapter 12. A. Dafni and C. O'Toole, Pollination syndromes in the Mediterranean: generalizations and peculiarities 125

Chapter 13. S.D. Johnson and W.J. Bond, Red flowers and butterfly pollination in the fynbos of South Africa 137

Seed dispersal

Chapter 14. E.J. Moll and B. McKenzie, Modes of dispersal of seeds in the Cape fynbos 151

Chapter 15. R.M. Cowling, S.M. Pierce, W.D. Stock and M. Cocks, Why are there so many myrmecochorous species in the Cape fynbos? 159

Index of key words 169

Author index 170

Systematic index 176

Preface

The Sixth International Conference on Mediterranean Climate ecosystems was held at Maleme (Crete), Greece, from September 23 to September 27, 1991. This conference had as its theme 'Plant-Animal Interactions in Mediterranean-type Ecosystems'. Most of the papers presented to that meeting have already been published (see Thanos, C.A. ed., 1992, Proceedings of the VI International Conference on Mediterranean Climate Ecosystems, Athens, 389 pp.). These 57 papers were all necessarily short. But the theme of plant–animal interactions was considered by the Organizing Committee to be so important to a fundamental understanding of the ecology of Mediterranean-climate ecosystems and to an enhanced management of those systems that various international research scientists were invited to prepare longer contributions on major aspects of the overall theme. The Book that follows represents the result of those invitations. All five regions of Mediterranean climate are represented – Chile, California, southern Australia and the Cape Province of South Africa, as well as the Mediterranean Basin itself.

Previous International Conferences on Mediterranean-type Ecosystems have had as their themes Convergence of Ecosystems and Biota (Chile, 1971), Fire and Fuel Management (California, 1977), Soil and Nutrients (South Africa, 1980), Resilience (Western Australia, 1984) and Time Scales and Water Stress (France, 1987). Already, planning for the next such Conference, to be held in Chile in 1994, is under way. The International Society of Mediterranean Ecologists (ISOMED) is an active one; these Conferences are organized under its *aegis*. It has been my privilege to be president of ISOMED for the period between the Fifth and the Sixth conferences. I congratulate the Organizing Committee for ensuring the undoubted success of the Sixth Conference at Crete, and especially Dr Margarita Arianoutsou for her role as Chairperson of that Committee. Further, I anticipate with pleasure the return of the next Conference to South America and to Chile, a country in which I have lived and worked.

Montpellier, February 1993 *Francesco di Castri*

Introduction

The Volume that follows comprises invited contributions on major aspects of Plant–Animal Interactions in Mediterranean Type Ecosystems, which was the subject of the Sixth International Conference on Mediterranean Climate Ecosystems, held in Crete (Greece), from September 23 to 27, 1991. The subject of plant–animal interactions was considered fundamental to the process of understanding the structure and the function of the Mediterranean type ecosystems and to their rational management. An attempt to present major new contributions on the issues of the overall theme of the Conference, beyond those already published in the Book of Proceedings (Thanos C.A. ed., 1992), seems warranted; that attempt follows.

All five regions of the world with a Mediterranean climate are represented: the Mediterranean Basin *sensu lato*, California, Chile, South Africa, Australia. Not all of the chapters have adopted the format of a review paper, although this was the original idea. Some contributors present recent research data, which they try to incorporate into general patterns, while others perform "challenging" interpretations to long established perspectives, calling for criticism.

The book is divided into six parts. In the first section, **Historical introduction**, the author tackles plant–animal interactions as described in the voluminous works of the founding parents of modern Biology, Aristotle and Theophrastus, the great philosophers of Greek Classical Antiquity. Besides the obvious tribute paid to the Conference location, this chapter reveals the important scientific accomplishments made by the two men during the era of 384–286 BC. **Community structure** includes chapters on the patterns of species richness and diversity of plants as they influence animal communities. As it is evident from almost all the chapters in this second section, plants that live in characteristic Mediterranean climates and that have evolved morphological and physiological adaptations, create the habitats to which the animals respond. The responses of animals to the high levels of temporal and spatial heterogeneity of these habitats are usually shown by a multiplicity of adaptations of the different groups of animals. The range of plant–animal interactions is enlarged by the innumerable soil microhabitats, caused either by the soil *per se* and/or by its invisible inhabitants. This is the subject of the third section presented under the title **Triangular relationships**. Plant morphology, habitat structure and seasonal fluctuations of available resources define species richness and structure in animal communities. The latter, in consequence, shape the form and sometimes the function and the vigour of the former. This is mostly accomplished through their feeding processes, **Herbivory**, which comprises the fourth part of the book. It is evident that no one book dealing with plant–animal interactions would be complete unless some information is presented on the important functions of the plants which are mostly animal-mediated, namely **Pollination** and **Seed dispersal**. These two subjects, forming the fifth and the last sections of the book respectively, reveal some of the peculiarities of the Mediterranean environment through the various pollination syndromes observed in plants of the Mediterranean Basin and those of the Cape fynbos as well as the specific mode of plant dispersal, myrmecochory, observed in the Cape flora.

These six sections on plant–animal interactions reflect the major trends in the direction of research. It is clear, however, that they do not cover the potential range of such kinds of relationships. Community structure of soil fauna, rodents and other mammals as affected by plant community structure, fire (and its direct or indirect effects on this bilateral relation), nutrient cycling and land use, are other issues that attract the attention of ISOMED scientists. These considerations become increasingly important in the light of increasing pressure on Mediterranean ecosystems, arising from recent economic development.

It is our wish that this volume will stimulate new and interesting research on the ecology of Mediterranean climate ecosystems. It is also our hope that the results of the research, which is either currently evolving or is scheduled for the future, will reinforce the movement for the greater conservation of these ecosystems, which have been one of the places for the origin of culture and science as we now know them.

Margarita Arianoutsou and Richard Groves, Athens and Montpellier *June 1993*

PART ONE

Historical introduction

CHAPTER 1

Aristotle and Theophrastus on plant-animal interactions

COSTAS A. THANOS
Institute of General Botany, University of Athens, Athens 15784, Greece

Key words: Aristotle, Theophrastus, zoology, botany, plant-animal interactions, biology

Abstract. Aristotle and Theophrastus, the last great philosophers and scientists of Greek Classical Antiquity, are the founding fathers of Zoology and Botany, respectively; they should also be honoured as the co-founders of Biology. They were close friends and life-long collaborators who evidently decided to pursue an organized study of the living world, probably in Lesbos at 344 BC (the landmark for the creation of the Science of Biology). The product of their division of labour, the voluminous zoological and botanical works of Aristotle and Theophrastus, respectively, were actually used contemporaneously as university textbooks by the students of the Lyceum.

Besides numerous comparisons and analogies, mostly on general issues, between animals and plants, both Aristotle and Theophrastus deal with various cases of plant-animal interactions, covering virtually all aspects of the field. Their scientific approach is notable, although the barriers to knowledge imposed by their era did not permit a significant contribution on issues like plant sex and pollination. Their important accomplishments on plant-animal interactions include herbivory and poisonous plants, plant pests and use of manure, insect-repellence and gall formation, fig caprification and apiculture, seed dispersal and seed infestation.

Introduction

Aristotle (384–322 BC) and Theophrastus (371–286 BC) should be considered as the last great philosophers of Greek Classical Antiquity. They represent, in particular, the culmination of the natural philosophy of the Ionian scientific tradition which was inaugurated on the Aegean coast of Asia Minor, several centuries earlier, and reached its climax on the opposite coast, at the brightest cradle of Ionian civilization, Athens. It should be borne in mind, however, that both Aristotle and Theophrastus were not Ionians but Macedonian (Dorian) and Aeolian, respectively; accordingly, they never became full citizens of Athens. During their lifetime they experienced the decline of the City-State system of Classical Greece and its eventual replacement by a more or less unified Greek State dominated by the northern Greeks (the Macedonians), under the leadership of Alexander the Great.

Aristotle was born at Stagira of Chalcidice (Macedonia) and his father Nicomachus was a physician at the court of Amyntas C′, father of Philip B′. At the age of 17, Aristotle moved to Athens where he studied and subsequently taught in Plato's Academy. Although by far the most brilliant among Plato's pupils, Aristotle was not appointed as the new director of the Academy after Plato's death (347 BC). Apparently as a result of his non-designation and with a team of colleagues and followers, he travelled to Assos of Troad (Asia Minor) and founded a new school there. Unfortunately, the venture came to an abrupt end three years later with the assassination of the patron of the school, Hermeias, the hegemon of Assos.

It is at that difficult moment in Aristotle's life that Theophrastus seems to have played a prominent role; probably at his suggestion (Morton, 1981), the two men moved to the nearby island of Lesbos, in the North Aegean Sea. Theophrastus (his original name was Tyrtamos) was born at the Lesbian town of Eresus and his father, Melas was a local fuller. It is probable that at an early age (ca. 355 BC) he went to

M. Arianoutsou and R.H. Groves, Plant-Animal Interactions in Mediterranean-Type Ecosystems, 3–11, 1994.

Athens and enrolled in the Academy, where he was acquainted and associated with Aristotle. According to Morton (1981), Aristotle evidently took to the highly intelligent, industrious and good-natured young man, who became his close friend and life-long collaborator.

After an apparent rest at Lesbos for two years (344–342 BC), Aristotle received an invitation by Philip B′, king of Macedonia, to serve as a tutor for his teen-aged son, the future Alexander C′ the Great. So Aristotle returned to his native country where he spent several years at the royal court, in the capital Pella, teaching the young prince 'το ευ ζην' (in free translation 'the quality of life'). Although there is no direct information, it seems highly probable that during his sojourn in Pella and afterwards in Stagira (when Alexander came of age), Aristotle was accompanied by Theophrastus.

In 335 BC, immediately after the struggle over domination of Greece was finally decided in favour of Macedonia, Aristotle returned to Athens, together with Theophrastus, and founded the Lyceum. After Alexander's death (323 BC), Aristotle had to flee Athens for Chalcis, Euboea where he died a few months later. Theophrastus became the second director of the Lyceum and the Peripatetic School (the name of the school was derived from the habit of lecturing while strolling around the gardens of the Lyceum) which reached its apogee of success during his 37 years of administration. The function of this institution was to train the leaders, officials and experts of the new era (Morton, 1981). Like its rival Academy, the Lyceum was a true University of its epoch: an up-to-date curriculum emphasizing the observational sciences, numerous lecturers, as many as two thousand students and a spacious, well designed campus with buildings and open-air facilities, a very important library, a museum and the first botanical garden.

Aristotle is generally considered one of the greatest Ancient Greek philosophers and during his rather short lifespan he wrote on virtually everything (in possibly as many as 400 works). Besides his philosophical treatises, he has left a number of voluminous works on natural history, the most important among them being the following (all of them fortunately extant): *Περι τα τωα ιστοριαι*, *HA* (Historia animalium, History of Animals, in 10 Books), *Περι ζωων μοριων*, *PA* (De partibus animalium, Parts of Animals, in 4 Books), *Περι ζωων κινησεως*, *MA* (De motu animalium, On the Movement of Animals, 1 Book), *Περι πορειας ζωων*, *IA* (De incessu animalium, Progression of Animals, 1 Book), *Περι ζωων γενεσεως*, *GA* (De generatione animalium, Generation of Animals, in 5 Books). The first work could be compared to what is currently considered a General Zoology text, whilst the second is the earliest treatise on Animal Physiology (the latter works covering more specialized fields). The also extant work *Περι φυτων* (De Plantis, On Plants, in 2 Books) that is included in Aristotle's minor works, is definitely not by him. It was almost certainly written more than 3 centuries later by Nicolaus of Damascus (lst century BC) and according to Morton (1981) reflects the level to which Peripatetic science was later reduced.

Theophrastus, though not admired as a major philosopher (of the stature of Aristotle, Plato or Democritus), was also a voluminous writer and is credited by Diogenes Laertius (3rd century AD) with 227 treatises. Apart from his well known Characters, his only other extant works are: *Περι φυτων ιστοριας* (Historia Plantarum, Enquiry into Plants, *HP*, in 9 Books) and *Περι φυτων αιτιων* (De Causis Plantarum, Causes of Plants, *CP*, in 6 Books). These works are the first, truly scientific botanical writings and correspond roughly to modern textbooks of General Botany and Plant Physiology, respectively. Among his non-extant works, 6 Books on the behaviour of animals are also included.

Peck (in his 1965 Introduction to *HA*) discusses all the relevant bibliography (in particular Thompson, 1910) concerning the dates of the treatise and concludes that Aristotle's natural history studies were carried out, or mainly carried out, between his two periods of residence in Athens, and especially during the 2-year stay at Lesbos (344–342). A similar conclusion is reached by Morton (1981) who suggests that Aristotle had not studied animals systematically until 344 BC when he moved to Lesbos. Kiortsis (1989) cites as writing dates the following: for *HA* 347–342 BC, for *PA* 330 BC and for *GA* 330–322 BC. Mitropoulos (in his Introduction to *HA*) believes that Aristotle's zoological works have been written in collaboration with several colleagues and disciples (Theophrastus, Strato, Eudemus and other Peripatetics). In his opinion, Theophrastus may have contributed much and, in particular, the spurious 10th Book of *HA* may belong exclusively to him. On the other hand, Balme (in his 1988 Introduction to the third volume of *HA*) concludes that there seems to be no compelling reason to believe that *HA* was written before the other biological treatises. In his opinion, all the available evidence suggests that Aristotle wrote *HA* I–IX as a study of animal differentiae; he collected the data initially from other treatises and then proceeded to complete the study from new reports, a process which was still unfinished at his death. The likeliest period for the bulk of his work is his stay at Lesbos (344–342 BC) and subsequent years (until 336 BC). Such a suggestion of course conflicts with the assumption that the *HA* was the collection of data which were to become the subject of further investigation in the other treatises. According to Balme, even Book X of *HA* is a genuine work of Aristotle, which nevertheless would possibly not belong to *HA* (it is probably the book listed in older catalogues as 'On failure to generate'). Concerning the dates of the works of Theophrastus, there exists only a suggestion by Morton (1981) that they were written in their final version around 300 BC. Hort (in his 1916 Introduction to *HP*) notes that the style of Theophrastus in his botanical works suggests that, as in the case of Aristotle, what we possess consists of notes for lectures or notes taken from lectures; there is no literary charm while the sentences are mostly compressed and highly elliptical, to the point sometimes of obscurity.

In my opinion, 344 BC may constitute the landmark for the creation of the Science of Biology in general and of its main constituents, Zoology and Botany, in particular (Thanos, 1992). A critical mass of technical and social changes in the Greek World (explicitly illustrated by Morton, 1981) motivated Aristotle and Theophrastus and resulted in their joint decision to pursue an organized study of the living world. Therefore, a rough division of labour was mutually agreed, Aristotle choosing animals and Theophrastus plants as their respective fields of interest. It should be no surprise that this decision was taken in the charming natural environment of Lesbos. As well, the origins of the two men must have played a role in that decision. Aristotle's father was a physician and he himself had some medical knowledge; Theophrastus, on the other hand, was closer to agriculture and forestry and evidently was aware, from his father, of many technical aspects of handling clothes and leather. I assume that the outlines of their major biological treatises had been thoroughly discussed and worked out during their stay in Lesbos; the bulk of their work would be already completed within the following decade, just in time for the inauguration of their Lyceum. If we accept that all these natural history books are simply University textbooks for the use of the Lyceum students, it is obvious that such treatises would continually be updated and corrected (and this would account partially for the confusion about their dates of composition, described previously). Therefore it may be deduced that concerning their biological works, the productive period for Aristotle was during his forties whilst for Theophrastus, it was during his thirties and forties. An interesting and relevant point is that although both pay full tribute to the earlier natural philosophers and naturalists (whom they cite in many instances: Alcmaeon, Anaxagoras, Empedocles, Democritus, Hippon, Menestor, Androtion, Chartodras, Diogenes of Apollonia, Cleidemus, Androkydes, Thrasyas of Mantineia, Leophanes) they never cite each other. Aristotle is never mentioned by name in Theophrastus although he is mostly prominent throughout *HP* and *CP*; the same holds true for Aristotle's works. The reason is that during this particular period it was considered good form not to mention a contemporary by name; this is, incidentally, solid proof that Theophrastus had already written the core of his works while Aristotle was still alive.

Aristotle is generally acknowledged nowadays as the founder of Biology in general and of Zoology in particular (e.g. Kiortsis, 1989). Theophrastus, on the other hand, although an extraordinary scientist, has only recently been recognized, internationally, as the founder of the science of Botany (e.g. Morton, 1981; Evenari, 1984). In my opinion both Aristotle and Theophrastus should be considered the co-founders of Biology.

The decision to split their study for 'academic' reasons may explain why both refrained from penetrating, in their writings, into each other's specific field. Nevertheless, overlapping and casual references to comparisons and interactions between plants and animals were obviously inevitable. Along with the discourse of Aristotle-Theophrastus on interaction and integration, the principal aim of this study has been the compilation and the critical analysis of the passages where animals and plants are cited in common. For this reason the previously mentioned major biological works of both men have been appropriately screened.

Textual references to plant-animal interactions

Aristotle, particularly in *GA*, and Theophrastus to a lesser extent are repeatedly attempting comparisons and analogies between animals and plants, though mainly on basic or general issues; as, for instance, in their discussions of general form and basic functions such as growth, nutrition and reproduction. According to Aristotle, plants are living creatures as well, but of a 'lower' level. Nature proceeds from the inanimate to the animals by small steps; the first step is plant life: it seems alive compared to inanimate things and inanimate compared to the animals (*HA* 588b). Plants lack locomotion (*PA* 656a) and have no power of sensation, or 'sensory soul' (like sea-squirts, sea-lungs and sponges, *PA* 681a, *GA* 741a; during sleep animals live like plants, *GA* 778b-779a). Nevertheless, in both plants and animals the principles ('souls') of growth, nutrition and generation are present (*GA* 735a, 740b). Plants do not have any excrement (*PA* 650a, 655b; similar to lower animals like sea anemones, *HA* 531b and ascidians, *PA* 681a), do not have real sexes (*GA* 731a), do not impregnate (like testaceans, *HA* 538a), accomplish their reproduction according to seasons (similar to certain sea animals: testaceans, sea-squirts, sea anemones and sponges, *HA* 588b) and can survive when dissected (like certain insects, *PA* 682b). A well-known assumption of Aristotle is that plants are like animals upside down (*PA* 650a); plants, like all living things have a superior and an inferior part but their superior part is in an inferior position (*IA* 705a-b, 706b). Because the superior part of plants is the roots which have the character and value of the mouth and head, the seed is the opposite, being produced at the top (*PA* 686b). After discussing similar subjects (*HP* 1.1.1.–1.2.6.), Theophrastus concludes: 'We should not expect to find in plants a complete correspondence with animals' (*HP* 1.1.3.), 'since plants, in contrast to animals, have no behaviour or activities' (*HP* 1.1.1.). Both Aristotle and Theophrastus draw the analogy of the leaf shedding habit in plants to the shedding of horns in stags, feathers in hibernating birds and hair in four-footed animals (*GA* 783b; *HP* 1.1.3.). Aristotle also compares blood vessels to the veins of broad leaves (*PA* 668a) and the umbilical cord through which the embryo receives its nourishment to the analogous structure (funicle) that connects the developing seed with the pericarp, as is nicely observed in pods (*GA* 752a, 753b).

Although Theophrastus was emphatically teaching botany instead of providing a local flora or a treatise on crop cultivation, the great economic importance agriculture had already gained during his time is obviously reflected to a certain degree in his works. Therefore, in numerous passages one finds discussions about manuring, particularly that concerning the effect of the various animal sources of manure to the growth of plants of various important crops (e.g. *HP* 2.2.11., 2.6.3., 2.7.3.–4., 6.7.6., 7.5.1., 8.7.7.; *CP* 3.6.1.–2., 3.9.1.–5., 3.17.5., 5.15.2.–3.). In addition, there are numerous references to animal pests and animal-caused diseases of plants, particularly crops (e.g. *HP* 3.12.8., 4.14.1.–10., 5.4.4.–5., 7.5.4., 8.10.1., 8.10.4.; *CP* 2.11.6., 3.22.3.–6., 4.14.4., 4.15.4., 4.16.1.–2., 5.9.3.–5., 5.10.1., 5.10.3., 5.10.5., 5.17.6.–7.). Considerable attention is devoted to seeds that are consumed by the larvae of beetles – thought to be produced by the seed itself – (e.g. *HP* 7.5.6., 8.10.5.; CP 5.18.1.–2.); this is a common case in legume seeds which are infested by larvae (today identified as the larvae of bruchid beetles), with the exception of chick pea (*Cicer arietinum*), bitter vetch (*Vicia ervilia*) and lupin (*Lupinus albus*) seeds which do not engender any creatures (*HP* 8.11.2.; *CP* 4.2.2.). In the case of chick pea, Theophrastus argues that it is a particular saltiness in the seed coat that prevents infestation (*CP* 6.10.6.).

As an obvious result of the great importance of galls as a tannin source, Theophrastus deals with them in detail and describes the external morphology of ten different types of galls produced by oak trees (*Quercus* spp., *HP* 3.7.4.–5.). He further reports several additional ones occurring in kermes-oak (*Quercus coccifera*, the well known scarlet 'berry' gall, *HP* 3.7.3., 3.16.1.), terebinth (*Pistacia terebinthus*, *HP* 3.15.4.), elm (*Ulmus glabra*, *HP* 3.7.3.) and laurel, in particular the male tree (*Laurus nobilis*, *HP* 3.7.3.). Although the general belief of his time was that the galls were formations of the plant itself, Theophrastus had noticed the presence of insects (resembling mosquitoes and flies, respectively) within the hollow bag-like gall of terebinth (*HP* 3.15.4.) and in the transparent, watery gall of the leaf rib of oak (*HP* 3.7.5.).

In discussing herbivory, Theophrastus makes a strong point that no general rule can be reached (*HP* 1.12.4.). Some parts of the plant may be edible and others inedible (CP 6.12.9.–11.); also some animals seem to prefer the tender parts while others prefer dry parts (*CP* 6.12.12.). He further states that it is usual that leaves are not edible while the fruits of the same plant can be consumed by both humans and animals. Less usual is the case of lime (*Tilia europaea*) with edible leaves but inedible fruits (*HP* 1.12.4.). Some plants are not touched by animals when they are green but are edible only when dried (after the sun has eliminated the 'bitterness'), as with sesame (*Sesamum indicum*), lupin (*Lupinus albus*) and possibly hedge-mustard (*Sisymbrium polyceratium*) and red-topped sage (*Salvia viridis*) (*HP* 8.7.3.; *CP* 6.12.12.). Another interesting observation is that animals find legumes a pleasure to digest (*CP* 4.9.1.), whilst legumes and fruits are part of the diet of the bear (*HA* 594b). According to Aristotle, sheep and goats are herbage eaters but, when foraging, the sheep graze intensively and stay in one place, while the goats quickly move on and only browse the tops; sheep are fattened on young olive shoots, wild olive (*Olea europaea* ssp. *oleaster*), tare (*Vicia sativa*) and any kind of brand (*HA* 596a). Cattle eat both grain and herbage, but are fattened on legumes such as bitter vetch (*Vicia ervilia*) and broad beans (*Vicia faba*). Horses, mules and asses eat grain and herbage (*HA* 595b); pigs are most inclined to eat roots and they are fattened on barley, millet, figs, acorns, wild pears and cucumbers (*HA* 595a). Some pasture species, such as lucerne (*Medicago sativa*), cause a failure of milk production, especially in ruminants; other pasture species, like cytisus (probably *Medicago arborea*) and bitter vetch (*Vicia ervilia*), increase the milk, although cytisus, when in bloom, causes burning and bitter vetch makes parturition more difficult (*HA* 522b).

Theophrastus mentions several examples of specific plants that produce toxic compounds and which may cause poisoning or death to the animals that might consume the particular plant parts. Examples furnished include black hellebore (*Helleborus cyclophyllus*), fatal to horses, cattle and pigs (*HP* 9.10.2.), the deadly root of wolf's bane (*Aconitum* sp.) which is not touched by sheep or other animals (*HP* 9.16.4.), the leaf and the fruit of the spindle-tree (*Euonymus europaeus*), fatal to both sheep and especially to goats (*HP* 3.18.13.), the leaf of yew (*Taxus baccata*), fatal to beasts of burden but not to ruminants (whilst its red 'fruit' is sweet and harmless to humans, *HP* 3.10.2.). Even three exotic plants (*Scorodosma foetida*, *HP* 4.4.12.; *Nerium odorum* and an unnamed plant, *HP* 4.4.13.) are cited as toxic. Aristotle also provides certain peculiar accounts: when a turtle has eaten some of a viper it eats oregano as well; when a weasel fights a snake it eats rue (*Ruta graveolens*) for its smell is inimical to snakes; storks and other birds apply oregano to a wound caused by fighting; when a snake eats fruit, it swallows the juice of 'bitterwort' (*HA* 612a) (vipers are also said to take rue after eating garlic, *CP* 6.4.7.). An interesting example is the highly poisonous root of *Thapsia garganica* which was never touched by the cattle indigenous to Attica, where this plant was particularly abundant. Imported cattle, on the other hand, fed on it and perished of diarrhoea (*HP* 9.20.3.). Fish are killed by the juice of black mullein (*Verbascum sinuatum*); hence people poison them in rivers and lakes whilst Phoenicians even used this poison in the sea (*HA* 602b). The Cretan dittany (*Origanum dictamnus*) is a fine example of healing properties. The plant is described by Theophrastus as being endemic to Crete and useful for many purposes, especially against difficult labour in women (*HP* 9.16.1.). The plant is both rare and with a very narrow distribution, attributable to the fact that goats are fond of it and graze it out. Theophrastus (*HP* 9.16.1.) adds also, somewhat sceptically though, the story of the arrow (cited by Aristotle as well, *HA* 612a) according to which a wounded wild goat (*Capra aegagrus cretica*) seeks to eat dittany and as a result the arrow drops off. This impressive story inspired the Flemish engraver Dapper, who illustrated it in 1703 (Baumann, 1982). In a comparable case the leaves and the stalk of silphium (*Ferula tingitana*) are said to be pleasant eating to sheep; it is also said that when a sick sheep is driven to graze in the silphium district it is quickly cured (*HA* 6.3.6.).

Another chapter concerns the repellant action of certain plants or plant products. In particular, hulwort (*Teucrium polium*) is good against moth in clothes (*HP* 1.10.4.). Furthermore, all insects find olive oil oppressive, for they avoid the mere smell of it (*CP* 6.5.3.) due to its pungency, just as with oregano (*Origanum* spp.) and the like (evidently, other aromatic labiates) (*CP* 6.5.4.). Ants will be made to abandon their nests if the entrances are sprinkled with oregano whilst most animals will flee if gum of storax (*Styrax officinalis*) is burnt (*HA* 534b). A strong insect-repellant action is reported for both species of *Inula* (*I. graveolens* and *I. viscosa*, the former – considered the 'female' kind – being more pungent, *HP* 6.2.6.); in addition, octopuses hold on so fast to the rocks that they cannot be pulled off unless they smell fleabane (*I. viscosa*) (*HA* 534b). As an example of mutually beneficial 'collaboration' radishes (*Raphanus sativus*) are reported to be interplanted with bitter vetch (*Vicia ervilia*) in order to prevent the latter from being eaten by flea-spiders (*HP* 7.5.4.; *CP* 2.18.1., 3.10.3.).

In the field of pollination one has to admit that no great contribution to our knowledge could be expected during Theophrastus' era. The real nature of flowers as sexual organs eluded Aristotle and Theophrastus. This comes as no surprise, since plant sex was suggested only two millennia later, in 1672, by the English physician Thomas Millington and the first experimental proof was furnished subsequently, in 1694, by Rudolf Jacob Camerer, a German professor of medicine at Tübingen (Bristow, 1980). Nevertheless, Aristotle and Theophrastus unconsciously distinguished the two sexes in certain dioecious plants on the basis of their fruiting ability, the 'female' being of course the fruit-bearing type; this same discrimination was generally extended on various domesticated plants as opposed to their wild relatives (e.g. olive-wild olive and fig-caprifig, *GA* 715b). Both Aristotle and Theophrastus believe that plants have no real sexes; Aristotle, in particular, is puzzled by the Testacea for which he states in several instances that having no sexes (and being sessile as well) they stand midway between plants and animals (*HA* 537b; GA 715b, 731b, 761a,b). According to Theophrastus the nature of seeds is close to that of eggs since they both contain in themselves a certain amount of food which is consumed with 'birth' (*CP* 1.7.1.). Aristotle similarly believes that in the living creatures where male and female are not separate, the 'seed' is as it were a foetus (*GA* 728b); he further states that animals with separate male and female parts seem to be just like divided plants: as though you were to pull a plant to pieces when it was bearing its seed and separate the male and female present in it (*GA* 731a). Thus things are alive by virtue of having in them a share of the male and of the female, and that is why even plants have life (*GA* 732a).

The case of the date palm (*Phoenix dactylifera*) has been well known since at least 1500 BC, as illustrated in the bas-reliefs of Nimrud in Mesopotamia (Meeuse & Morris, 1984). There is strong evidence that even the ancient Assyrians were familiar with the practice of artificial pollination of date palm. Theophrastus, after describing this procedure (*HP* 2.8.4.; *CP* 3.18.1.), advances further to the point of comparing (*CP* 2.9.15.) the dusting of the female flowers by the male inflorescence to what was observed with fish, when the male sprinkles his milt on the eggs as they are laid (*GA* 755b). Another very interesting case is fig caprification, the fig tree (*Ficus carica*) being a close companion of man for many millennia in the Eastern Mediterranean area. Concerning fig caprification, Theophrastus devotes two lengthy and quite exhaustive passages (*HP* 2.8.1.–3.; *CP* 2.9.5.–14.) and Aristotle a short but comprehensive one (*HA* 557b). In certain cultivated fig varieties (considered the 'female' fig trees), caprification is the necessary procedure to promote proper maturation of the syconium, the complex fruit of fig. Thus wild figs (produced by so-called 'male' fig trees) were hung on the cultivated fig tree or wild trees were planted on eminences near the fig orchard to ensure the insects an easy flight down wind (*CP* 2.9.5.). For it is particular insects that, after having grown inside the developing seeds of the wild figs, will pierce the pericarp tissues, to seek another fig to get in (*HP* 2.8.1.; *CP* 2.9.5.) and spend the rest of their life (having carried the pollen from the first to the latter while at the same time laying their eggs). These insects are the 'psenes' of Theophrastus, the presently called fig wasps, Agaontidae (the particular one being *Blastophaga psenes*). Theophrastus also describes another type of wasp which never gets out of the fig and is sluggish like a drone (*HP* 2.8.2.). These are actually the wingless male wasps that fertilize the females before the latter make their trip to their second fig-host. (For a modern account of fig biology the reader is referred to Galil, 1977.)

Aristotle appears to be in possession of both accurate and deep knowledge in regard to apiculture, although he was obviously ignorant of the role of pollen and nectar. After sceptically mentioning that, according to some, bees are generated spontaneously from the flowers of a broom, reed or olive (*HA* 553a), he correlates a heavy crop of olives with frequent bee swarming (*HA* 553b) since wax collecting had been observed on the olive trees (*HA* 624b). It is from the flower of thyme (*Coridothymus capitatus*) that the bees get the honey and according to the abundance of its flowering the beekeepers can forecast a rich or a poor yield (*HP* 6.2.3.). Thyme honey is outstanding for its sweetness and consistency and can be distinguished immediately (*HA* 554a). Besides thyme as a source of food (*HA* 626b), during blossom periods bees collect from the following plants: 'atraktyllis' (*Carthamus* sp.), honey-lotus (*Trigonella graeca*), asphodel (*Asphodelus* sp.), myrtle (*Myrtus communis*), 'phleos' (*Saccharum* sp.), chaste-tree (*Vitex agnus-castus*) and broom (*Spartium junceum*) (*HA* 627a). It is also beneficial to plant around the hives wild pears (*Pyrus amygdaliformis*), almond (*Prunus amygdalus*), myrtle (*Myrtus communis*), broad beans (*Vicia faba*), lucerne (*Medicago sativa*), Syrian grass (probably a legume), winged vetchling (*Lathyrus ochrus*), poppy (*Papaver rhoeas*) and herpyllus thyme (*Thymus* sp.) (*HA* 627b). In Pontus (Black Sea) there exist white bees which produce honey twice per month but only during the winter because they collect honey from the abundant ivy (*Hedera helix*) (*HA* 554b). Aristotle had noticed that on each individual flight the bee visits only plants of the same kind (*HA* 624a) and the wax (i.e. the pollen) is carried on the legs (*HA* 554a); the bees pick it up by scrabbling at the blossoms busily with their front feet and subsequently wiping it off to the middle and hind ones (*HA* 624a). The honey is gathered with their mouth from all flowers whose blossoms are in a calyx and from all others which contain sweetness (nectar?), without any injury caused to the fruit; finally the honey is vomited in the cells of the comb (*HA* 554a). Bees were observed to sicken when they worked on mildewed plants and the best honey was made from young wax (pollen) and 'moschos' (nectar?) (*HA* 626b). The comb comes from flowers and is sealed with tree gum whilst the honey is made from what falls from the air (*HA* 553b), an unfortunate conclusion due to the fact that no direct correlation with flowering could be drawn: the hives were found filled with honey within one or two days (and not for instance in the autumn, although there was blossom enough) (*HA* 553b).

In the field of seed and fruit dispersal, numerous accounts are furnished, especially by Theophrastus. The cormlets of the corn-flag (*Gladiolus segetum*) are found in the runs of moles (*Talpa europaea*) 'for this animal likes them and collects them' (*HP* 7.12.3.), this habit leading, unintentionally of course, to dispersal of the plant. A similar case of accidental dispersal and regeneration is in oaks (*Quercus* spp.), through the caching of acorns by jays (*Garrulus glandarius*) and other birds (*HA* 615b; *CP* 2.17.8.). An example of epizoochory is the fruit of goosegrass (*Galium aparine*) which sticks to clothes (*HP* 7.14.3.). Endozoochory is represented by ivy (*Hedera helix*) fruits reported to occur either as bitter or sweet and consumed by birds only in the latter state (*HP* 3.18.10.). An interesting example of removal of hardseededness by the passage through the digestive tract is furnished by the pods of *Colutea arborescens*: they are described as a wonderfully fattening food for sheep whilst best seed germination is obtained by using the sheep-droppings (*HP* 3.18.2.). A final fine example of zoochory is the case of the mistletoe; it is at the same time a masterly treatise by Theophrastus (*CP* 2.17.1.–10.), which, incidentally, has escaped the attention of mistletoe specialists such as Calder & Bernhardt (1983). In this discourse, Theophrastus, after having identified the two species occurring in Greece (*Loranthus europaeus* and *Viscum album*) (*CP* 2.17.1.), asks questions and provides answers concerning the peculiar habit of these plants not to grow on the ground but only on a host tree (*CP* 2.17.3.) of various species: *Abies* spp., *Pinus* spp., *Pistacia terebinthus* and *Quercus* spp. (*CP* 2.17.1.), even *Q. coccifera* (*HP* 3.16.1.). He concludes that mistletoes are dispersed by birds in the seed state since it is the birds that consume the mistletoe berries (*CP* 2.17.5., 2.17.8.); incidentally, Aristotle, in describing the three species of thrush he had observed, mentions that one of them (most probably the mistle thrush, *Turdus viscivorus*) eats only mistletoe berries (*HA* 617a). The seeds pass unharmed through the digestive tract of the bird and are able to establish their new seedlings only when the droppings happen to fall on a host plant (*CP* 2.17.5.).

Besides (or maybe due) to the rough division of labour into two main, scientific domains, most scattered references to plants in Aristotle and a considerable number of references to animals in Theophrastus are of a general type and follow a very consistent pattern of principled comparisons between animals and plants. The overall impression is the conception by both men of a single natural kingdom of living organisms, a profoundly scientific view derived from the naturalism of the Ionian philosophers and in marked contrast to the idealism of Plato (Morton, 1981).

The extent of the greatness of Aristotle and Theophrastus as scientists will be additionally illustrated by the striking passages that follow. According to Aristotle it is not the size of the body that determines the size of the brood; and it is not only among the animals that walk but also among those that fly and swim that the big ones produce few offspring and the small ones produce many. Similarly too it is not the biggest plants that bear the most fruit (*GA* 771b). Theophrastus also notes that plants with smaller seeds produce more of them and, similarly, certain animals (especially the oviparous and larviparous ones) bring forth small but numerous progeny (*CP* 4.15.2.). In prolific birds, nourishment is allocated to the semen; some fowls after having laid excessively, die. Similarly many trees wither away when they have borne an excessive amount of fruit since no nourishment is left for themselves. Annual plants (legumes, cereals) experience the same thing. Their kind produces a great deal of seed and they use up all their nourishment for seed. The birds and plants alike become exhausted (*GA* 750a). Theophrastus observes that the most prolific animals are the quickest to age and die whilst, similarly, the plants that age earlier are those that bear many crops and abundant fruit (*CP* 2.11.1.), or too large a crop (in both trees and annuals) (CP 2.11.2.–3.). These remarks should be considered as the earliest hints of *r*- and *K*-selection strategies as well as of the principle of allocation and reproductive effort. These important concepts have only recently been established in the fields of reproductive biology and evolution; in particular the rapid death of monocarpic plants following reproduction may simply be a consequence of exhaustion, because of excessive reproductive effort (for a comprehensive account of these concepts the reader is referred to Fenner, 1985).

A final point is that although Aristotle and Theophrastus were by any standards great philosophers and scientists, their works reflect to a great extent the overall attitude towards nature, at the end of Classical Antiquity, as well as the level of accumulated knowledge of Greek society in general at that time.

Acknowledgment

Thanks are due to Professor B.A. Kyrkos for useful comments and suggestions.

References

Aristotle. Historia Animalium, vol. I. & II. Peck, A.L. (translator, 1965 & 1970), vol. III. Balme, D.M. (translator, 1991), Harvard University Press – William Heinemann, Cambridge, Mass., London.

Aristotle. History of Animals. Mitropoulos, K. (translator

into modern Greek), Library of Greeks, vol. 126–130. Greek Publishing Organization, Athens.

Aristotle. Parts of Animals, Movements of Animals, Progression of Animals. Peck, A.L. & Forster, E.S. (translators, 1937), Harvard University Press – William Heinemann, Cambridge, Mass., London.

Aristotle. Parts of Animals. Mitropoulos, K. (translator into modern Greek), Library of Greeks, vol. 74–75. Greek Publishing Organization, Athens.

Aristotle. Generation of Animals. Peck, A.L. (translator, 1942), Harvard University Press – William Heinemann, Cambridge, Mass., London.

Aristotle. Generation of Animals. Mitropoulos, K. (translator into modern Greek), Library of Greeks, vol. 76–77. Greek Publishing Organization, Athens.

Baumann, H. 1982. Die Griechische Pflanzenwelt in Mythos, Kunst und Literatur. Hirmer Verlag, München.

Bristow, A. 1980. The Sex Life of Plants. New English Library, London.

Calder, M. & Bernhardt, P. (eds) 1983. The Biology of Mistletoes. Academic Press, Sydney.

Evenari, M. 1984. Seed physiology: its history from Antiquity to the beginning of the 20th century. Bot. Rev. 50: 119–142.

Fenner, M. 1985. Seed Ecology. Chapman and Hall, London and New York.

Galil, J. 1977. Fig biology. Endeavour 1: 52–56.

Kiortsis, B. 1989. Aristotle the founder of Biology. Biology, philosophy, science, perspectives (in Greek). Hellenic Society of Biological Sciences, Athens.

Meeuse, B. & Morris, S. 1984. The Sex Life of Flowers. Facts on File Publications, New York.

Morton, A.G. 1981. History of Botanical Science. Academic Press, London.

Thanos, C.A. 1992. Theophrastus on plant-animal interactions. In: Thanos, C.A. (ed.), Plant-animal interactions in Mediterranean-type ecosystems, 1–5. Proceedings of MEDECOS VI, Athens.

Theophrastus. Enquiry into plants, vol. I. & II. Hort, A.F. (translator, 1916 & 1926), William Heinemann – Harvard University Press, London, Cambridge, Mass.

Theophrastus. De causis plantarum, vol. I., II. & III. Einarson, B. & Link, G.K.K. (translators, 1976 & 1990), William Heinemann – Harvard University Press, London, Cambridge, Mass.

Thompson, D'Arcy W. 1910. The works of Aristotle translated, vol. IV. Oxford.

PART TWO

Community structure

CHAPTER 2

Species richness of vascular plants and vertebrates in relation to canopy productivity

RAYMOND L. SPECHT
Botany Department, The University of Queensland, St Lucia, Queensland 4072, Australia
Current address: 107 Central Avenue, St Lucia, Queensland 4067, Australia

Key words: community diversity, species diversity, vascular plants, vertebrates, solar radiation, primary productivity

Abstract. Three paradigms are proposed to explain the community-physiological processes operating to maintain species diversity in an ecosystem.

1. The species richness (number of vascular plant species per hectare) of Australian plant communities is related to the amount of incident solar radiation intercepted by both overstorey and understorey strata, and the annual shoot growth which results in these strata.
2. The species richness of the vascular plants (of different life-form, e.g. shrub, low shrub, perennial or annual herb) in the understorey will depend on the percentage of net photosynthates which is translocated to shoot and floral apices.
3. The species richness of amphibia, birds, mammals and snakes all appear to parallel the increase in the species richness of vascular plants along the climatic gradient from the semi-arid to the humid zone. Species richness of lizards decreases as the overstorey of the plant community becomes denser along the same climatic gradient.

The paradigms derived from the Australian data appear to be applicable to other Mediterranean-type ecosystems. Species diversity of the overstorey is invariably low in Mediterranean regions. If the overstorey regenerates rapidly from densely-packed rootstocks during post-fire succession, the resultant foliage canopy suppresses the growth and diversity of the understorey; the species diversity of vascular plants in these ecosystems is low, but species diversity of vertebrates appears to be unaffected.

The processes which determine the species richness of plant communities and associated vertebrates needs immediate research, to provide a basis for scientific management of ecosystems to maintain biological diversity.

Introduction

The concerted efforts of many ecologists interested in Mediterranean-type ecosystems (MTEs) enabled a Data Source Book to be published (Specht, 1988a). This volume summarised the many attributes of MTEs throughout the world. The species richness (alpha diversity) of plants, in 1, 10, 100 and 1000 m^2 quadrats, was summarised by Westman (1988), using the species-area curve of Hopkins (1955).

Information compiled in Specht (1988a) provides a ready summary of the variation in species richness which may be observed in MTEs throughout the world.

Mediterranean-type ecosystems in warm-temperate, southern Australia have evolved over the last five to ten million years from a subtropical Gondwanan vegetation which developed during the Late Cretaceous (Specht *et al.*, 1992). The onset of aridity, with falling temperatures during the Late Tertiary and the Quaternary, induced changes in the perhumid to humid, subtropical vegetation of southern Australia – the Mediterranean-climate ecosystems, which are found today in the arid to perhumid climatic zones, evolved. In order to understand the processes which determine species diversity in present-day MTEs, it is necessary to trace the changes which have occurred in the Early Tertiary vegetation of southern Australia (Specht & Dettmann, in press).

It would appear that most Mediterranean-climate ecosystems of the world have experienced the same evolutionary trends observed in Australia, developing from a warmer subtropical climate with rainfall uniformly distributed through-

M. Arianoutsou and R.H. Groves, Plant-Animal Interactions in Mediterranean-Type Ecosystems, 15–24, 1994.

out the year (Axelrod, 1973, 1975, 1989; Pons, 1981). The island continent of Australia, extending from tropical latitude 10° S to the cool temperate latitude 45°S, provides a range of present-day examples of ecosystems from which the MTEs of southern Australia have evolved (Specht & Dettmann, in press). Detailed observations on the growth and species diversity of ecosystems from tropical to warm temperate Australia have enabled some understanding of the processes which appear to control the species richness (alpha diversity) of ecosystems in the Mediterranean-climate of southern Australia (Specht, 1972, 1981, 1988b; Specht & Specht, 1989a,b,c; Specht *et al.*, 1990; Specht & Specht, 1993).

The results of these Australian studies may be summarised in the following paradigms (generalised models), which appear to have universal application to other Mediterranean-type climatic regions of the world.

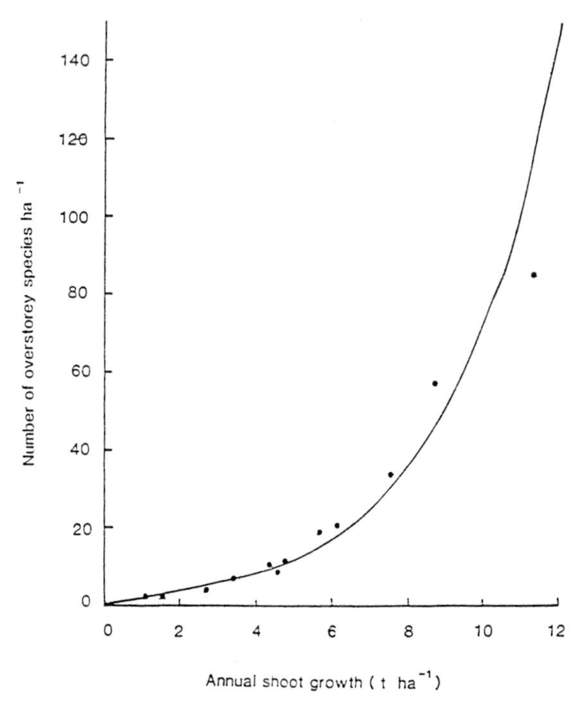

Fig. 2.1. Species richness (mean number of vascular species per unit area) of the overstorey of Australian plant communities in relation to the annual productivity of the foliage canopy (Specht & Specht, 1993).

Paradigms of species richness (alpha diversity) of plant communities

1. Species richness (number of vascular plant species per one hectare) of the overstorey of lowland plant communities is exponentially correlated with annual shoot growth (but not with the annual growth of stems) of the overstorey foliage canopy, which itself is influenced by solar radiation, water and nutrient stress (Fig. 2.1)
   a. Annual shoot growth of the overstorey (when water is not limiting) increases exponentially as incident solar radiation increases from the temperate to the tropical climatic regions (Specht & Specht, 1993). When stem-root respiration equals net photosynthesis of the foliage canopy, annual shoot growth of the overstorey tends to zero. This occurs at the latitudinal or altitudinal tree-line, when the mean annual temperature (of the snow/frost-free season) is less than 13 °C.
   b. Annual shoot growth of the overstorey (in any climatic region) increases exponentially from the arid to the perhumid climatic zone (Specht & Specht, 1993).
   c. Annual shoot growth of the overstorey (in any climatic region and zone) increases asymptotically from nutrient poor soils to nutrient rich soils (Fig. 2.2).
2. Species richness (number of vascular plant species per one hectare) of the understorey is related to the amount of incident solar radiation penetrating through gaps in the overstorey foliage canopy.
   a. Species richness of the understorey of the plant community (in the regeneration phase after disturbance) increases exponentially as incident solar radiation increases from the temperate to the tropical climatic regions (Specht & Specht, 1993).
   b. Species richness of the understorey decreases during the developmental phase of the plant community, as the amount of incident solar radiation penetrating through gaps in the overstorey foliage canopy decreases with ageing of the plant community (Fig. 2.3).

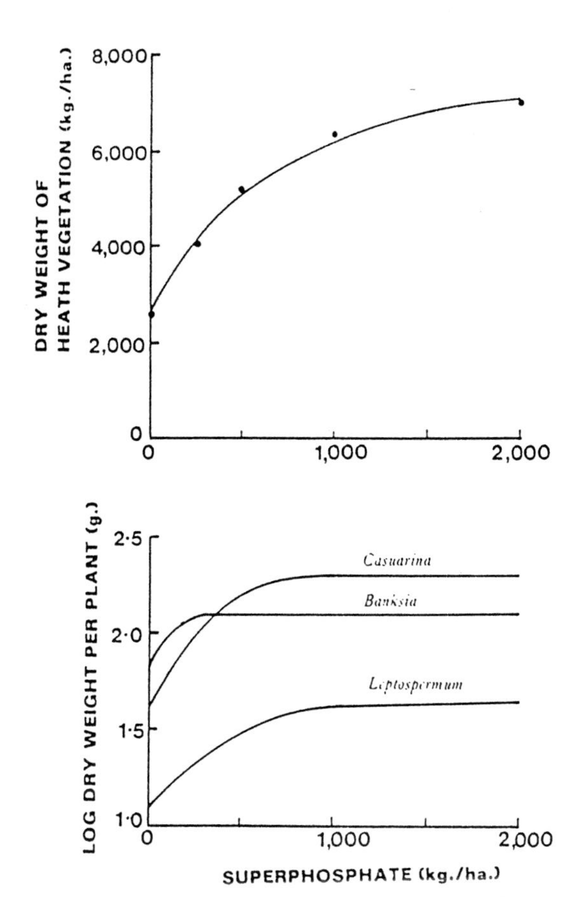

Fig. 2.2. Growth response of heathland on nutrient-poor soil to applied phosphatic fertilizer in the Mediterranean-type climate of south-eastern Australia (Specht, 1963; Heddle & Specht, 1975).

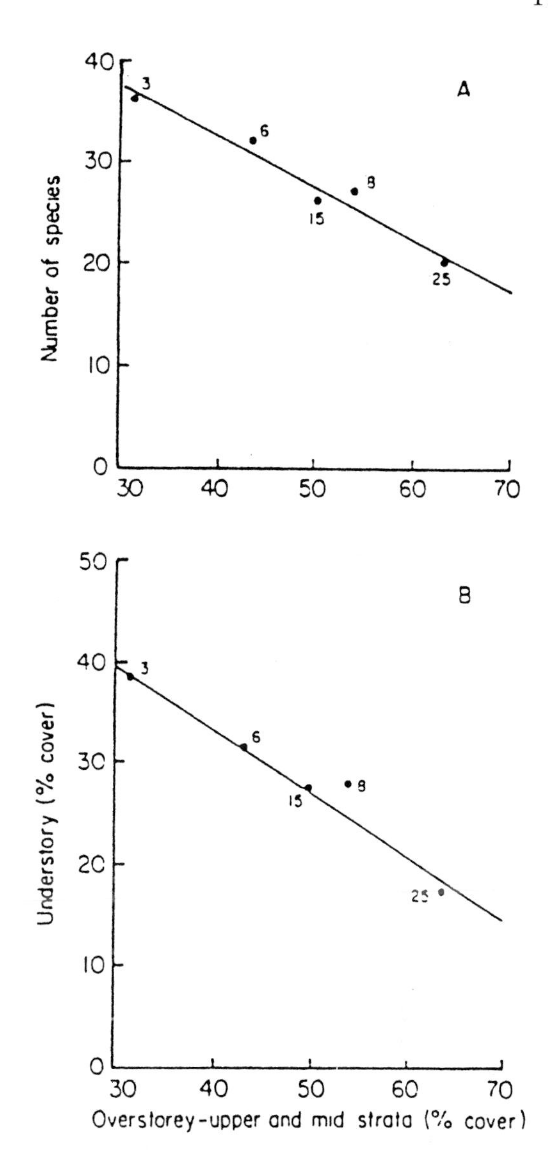

Fig. 2.3. The influence of the Foliage Projective Cover (%) of the overstorey on the species richness (number of vascular plants per unit area) of the understorey, as a Mediterranean-climate heathland regenerates after fire (Specht, 1963; Specht & Morgan, 1981; Specht & Specht, 1989c).

c. Species richness of the understorey decreases with an increase in the amount of incident solar radiation intercepted by the overstorey:
   i. As the angle of the sun increases from the tropics to the poles (Fig. 2.4).
   ii. As overstorey foliage cover is increased (relative to the norm):
      (i) With regeneration of the overstorey of the plant community (towards the norm) during secondary succession (see Fig. 2.3).
      (ii) With increased density of the overstorey of the mature plant community in the normal climatic sequence from the arid to the perhumid zone (Fig. 2.5).
      (iii) When the density of overstorey trees increases abruptly to produce a closed forest (Fig. 2.6), in oases where water is no longer limiting during any month (both in freely-

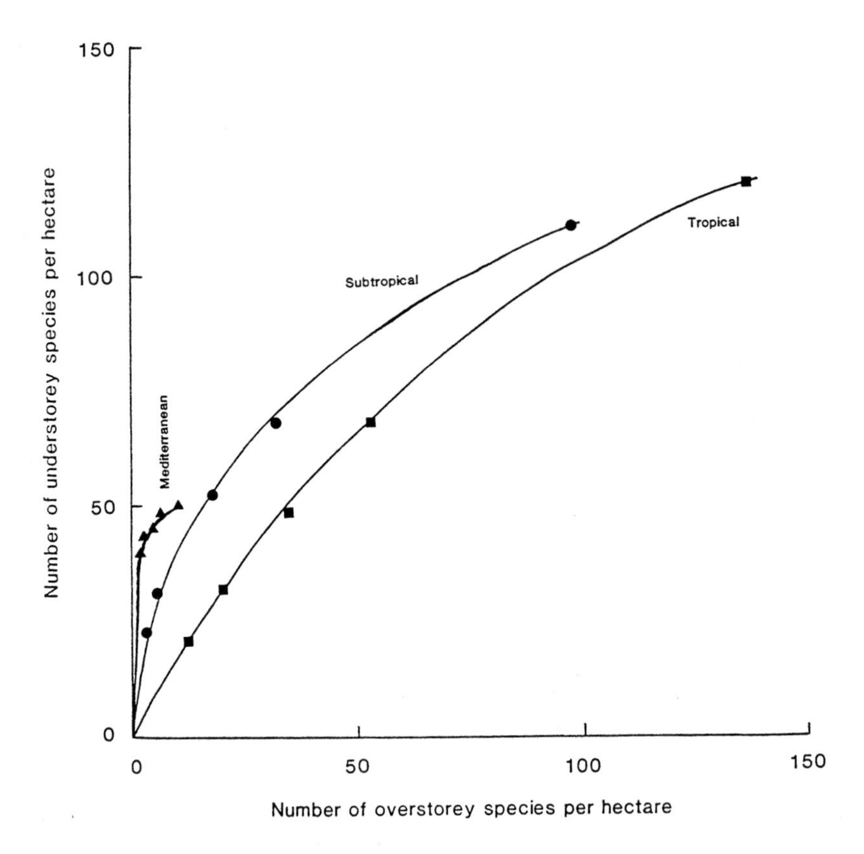

Fig. 2.4. Species richness (N, number of vascular species per hectare) of the understorey is plotted against the species richness of the overstorey, recorded in plant communities of the arid, semi-arid, subhumid, humid and perhumid climatic zones of warm temperate (Mediterranean), subtropical and tropical regions of Australia. Species richness of understorey and overstorey strata in the perhumid climatic zone from cool temperate to tropical Australia may be expressed by the following equation:

$$\ln(125 - \text{understorey N}) = 4.68 - 0.022 \text{ overstorey N}$$

drained and poorly-drained habitats).

(iv) With an increase in the density of long-lived rootstock-regenerators, e.g. *Quercus ilex* forest, *Quercus coccifera* garrigue, *Adenostoma-Ceanothus* chaparral. (Was this increase in density initiated during a previous wetter climatic cycle and maintained, even today, by the rootstock regenerators?) (see Fig. 2.6).

d. Species richness of the understorey is increased (relative to the norm) when optimal overstorey foliage cover fails to establish:

i. At the latitudinal and altitudinal tree-lines.

ii. In areas which contain a sequence of nutrient poor soils of increasing poverty (Fig. 2.7).

iii. As the soils become increasingly water-logged.

iv. As the surface layer of the soil becomes deeply cracked and seasonally desiccated.

3. Species richness (number of vascular plant species per one hectare) of the understorey is correlated with the life form (and seed production) of the understorey (Fig. 2.8). As the amount of respiring tissue of the understorey increases, the proportion of net photosynthate available for flower and seed production decreases, as does species richness of the understorey. In effect, the species richness of an understorey of therophytes is greater than that

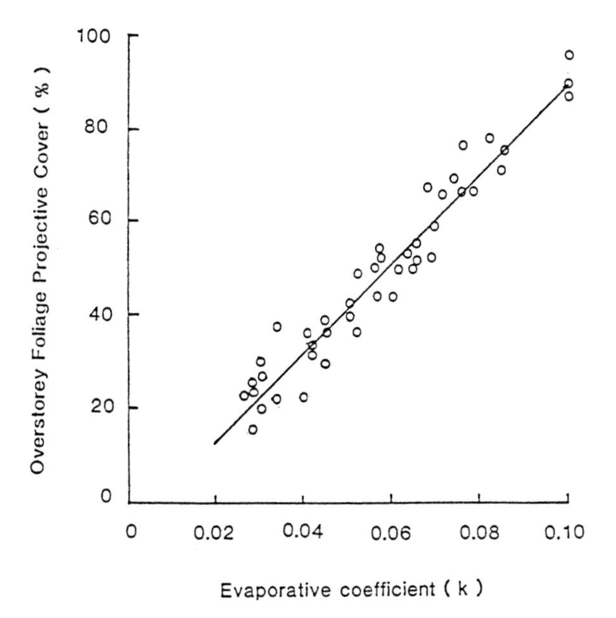

Fig. 2.5. Increase in the Foliage Projective Cover (%) of the canopy of mature overstoreys along the climatic gradient (evaporative coefficient) from the arid to the perhumid zone in warm temperate (mediterranean), subtropical and tropical regions of Australia (Specht, 1972, 1981).

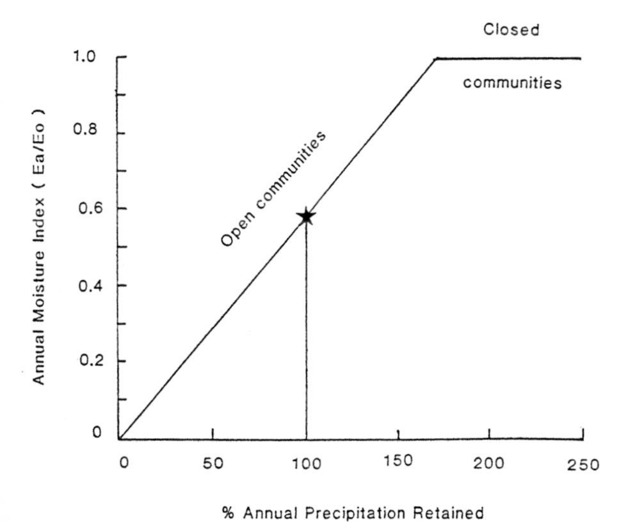

Fig. 2.6. Open communities change abruptly to closed communities (with Foliage Projective Cover 70–100%) when available water is non-limiting during every month of the year (Specht, 1972, 1981).

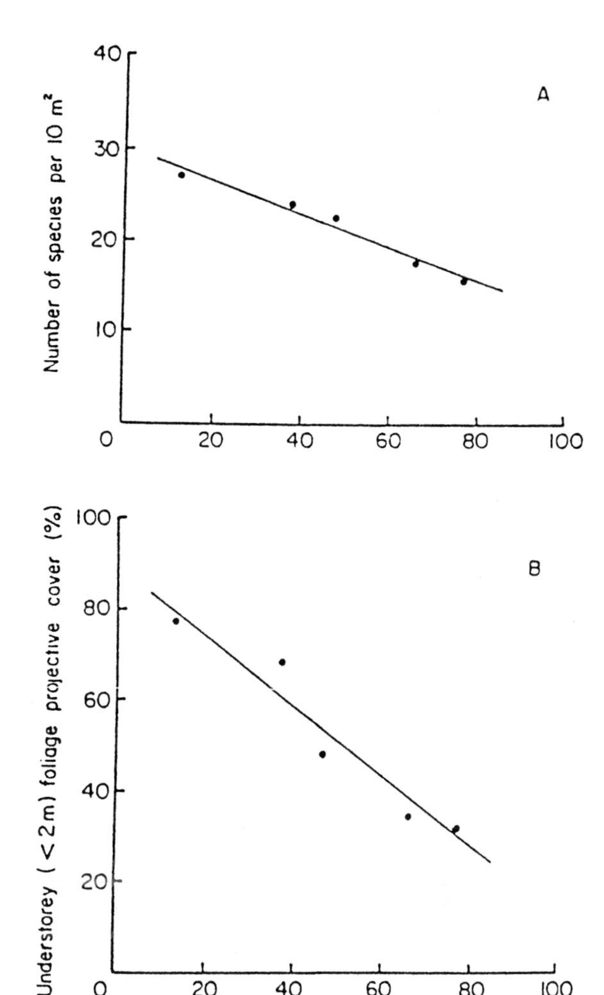

Fig. 2.7. The influence of Foliage Projective Cover of the overstorey on the species richness (number of vascular species per unit area) of the understorey, in the soil-nutrient gradient from heathland to heathy open-forest (Specht & Morgan, 1981).

of an understorey composed of hemicryptophytes and geophytes, and far greater than that of an understorey composed of chamaephytes and nanophanerophytes.

Paradigms of species richness (alpha diversity) of vertebrates

1. The number of species of resident amphibia, birds, mammals and snakes in an ecosystem

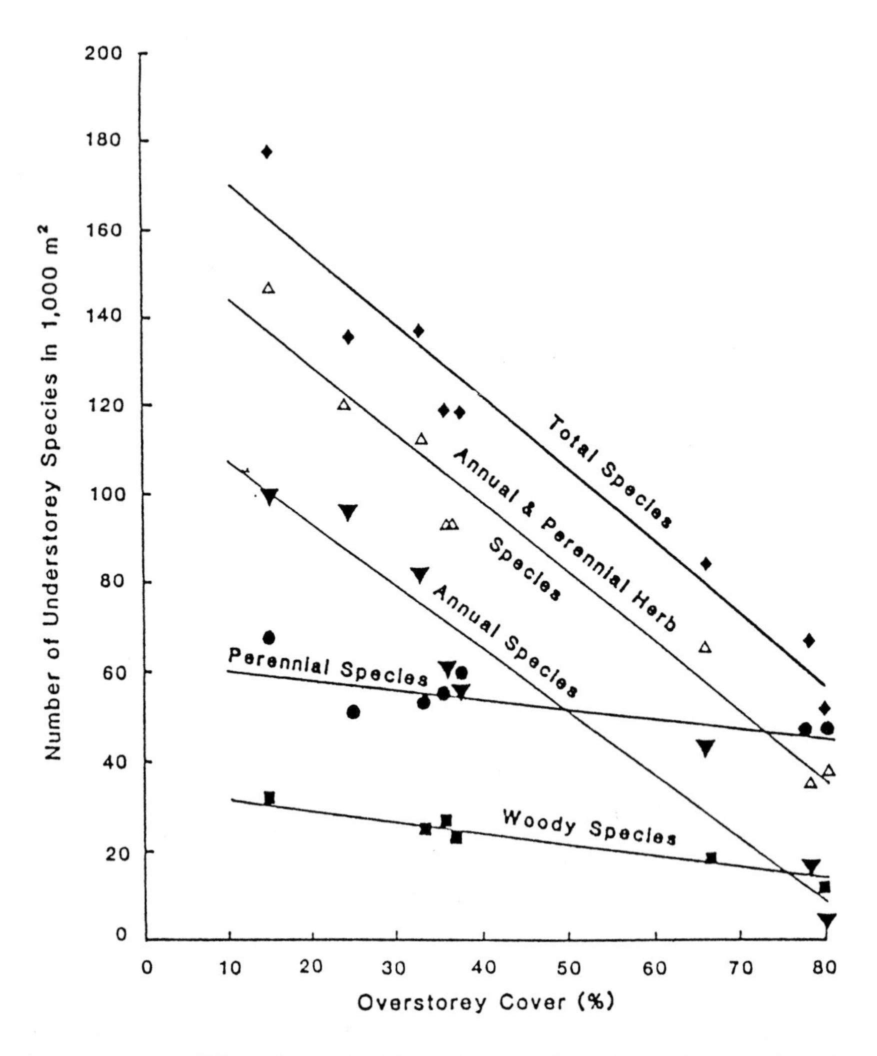

Fig. 2.8. The effect of overstorey cover (%) on the species richness (number of vascular species per unit area) of annual herbs, perennial grasses and geophytes, and woody shrubs in northern Israel (Specht *et al.*, 1990).

increases in the climatic gradient from the semi-arid to the humid zone:

a. As the annual shoot growth (the food source) of the plant community increases along the climatic gradient from the arid to the humid zone (Fig. 2.9).

b. As the diversity (species richness) of the plant community increases (Fig. 2.10).

c. As the structure of the plant community becomes more complex.

2. The number of species of resident lizards in an ecosystem decreases in the climatic gradient from the semi-arid to the humid zone.

a. As the amount of solar radiation reaching the understorey, through the overstorey gaps, decreases (Fig. 2.11).

3. The number of species of resident birds, lizards and mammals in the ecosystem is reduced when annual shoot growth (the food source) of the plant community is reduced, on soils extremely poor in mineral nutrients:

a. As soil nutrient levels decrease in a nutrient-gradient from heathy open-forest to heathland. In contrast, species richness of the plant community increases along the nutrient gradient as overstorey cover is reduced (Table 2.1).

b. In the lower soil nutrient levels of south-

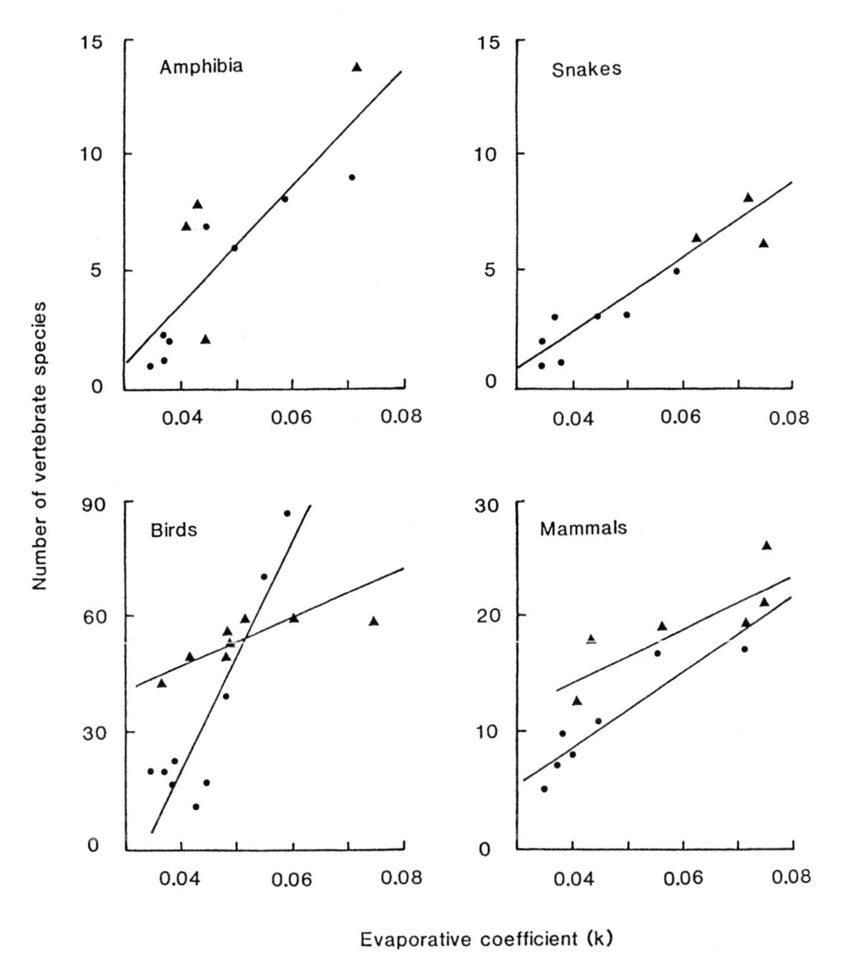

Fig. 2.9. Total number of amphibia, birds, mammals and snakes resident in plant communities in Mediterranean southwestern (●) and southeastern (▲) Australia (Catling, 1988), plotted along the climatic gradient from semi-arid (k = 0.035 – 0.045), to subhumid (k = 0.045 – 0.055), to humid (k = 0.055 – 0.075). Exceptionally high values of 'resident' birds recorded in mallee-broombush communities in southeastern Australia probably include many migratory species (Edmonds & Specht, 1981).

western in contrast with south-eastern Australia (see Figs. 2.9 and 2.11).

Paradigms of community richness (in 1° latitude × 1° longitude grid)

1. Community richness (per grid square), as defined by the classificatory program TWINSPAN (Hill, 1973; Specht *et al.*, in press), increases as the climate becomes more humid, from the arid to the perhumid zone (Fig. 2.12). (The humid/perhumid zone is usually centered on coastal highlands which uplift the prevailing weather systems).
2. Community richness (per grid square) increases (exponentially) as solar radiation increases from temperate to tropical climatic regions of Australia.
3. Community richness (per grid square) will depend on the degree of weathering of the landscape (gently-dissected pine-plain versus deeply-dissected highland).

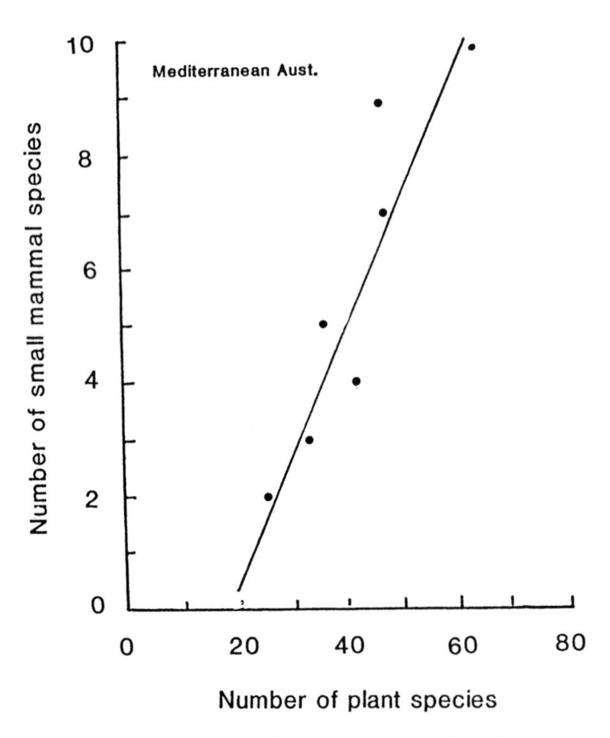

Fig. 2.10. Number of small mammals recorded in plant communities in Mediterranean southern Australia (Catling, 1988) plotted against the number of plant species recorded in the same ecosystems (corrected from Specht, 1988b).

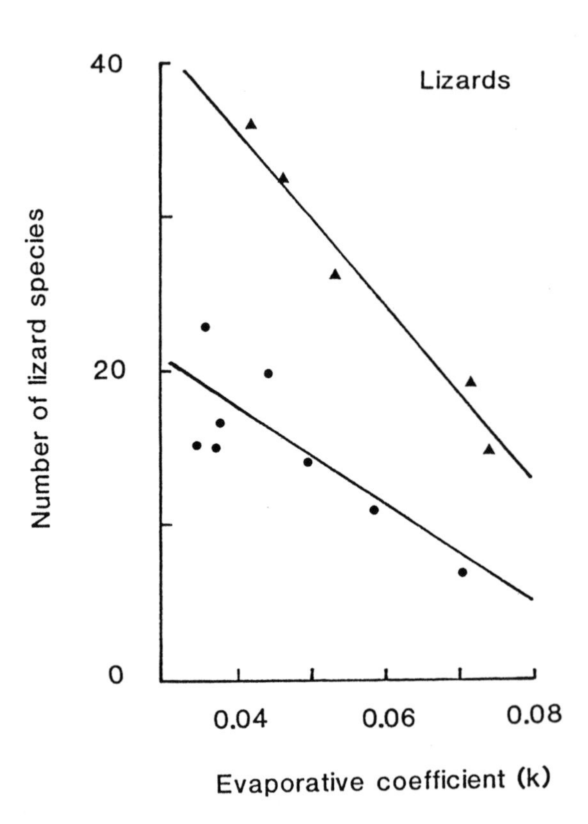

Fig. 2.11. Total number of species of lizards resident in plant communities in Mediterranean southwestern (●) and southeastern (▲) Australia (Catling, 1988), plotted along the climatic gradient from semi-arid (k = 0.035 – 0.045), to subhumid (k = 0.045 – 0.055), to humid (k = 0.055 – 0.075).

Discussion

Recent studies on community-physiological processes, which appear to control the species richness (alpha diversity) of ecosystems, have enabled a series of paradigms (generalised models) to be developed concerning

1. Species diversity of vascular plants.
2. Species diversity of vertebrates.
3. Community diversity in 1° latitude × 1° longitude grid.

Species richness (number of vascular plant species per hectare) of overstorey and understorey strata within a plant community is exponentially correlated with the amount of shoot growth produced annually per hectare by each stratum.

In general, the number of species of endothermic vertebrates in a series of ecosystems increased linearly as the number of species of vascular plants per hectare increases. This linear correlation between species of vertebrates and species of vascular plants, and the exponential correlation between species richness of vascular plants and annual shoot growth per hectare, suggests that species richness of vertebrates is also controlled by the amount of energy fixed as shoot growth in a form readily available to herbivores.

Table 2.1. Total number of mammals and birds recorded in adjacent areas of heathy vegetation, with and without *Eucalyptus* trees, for two regions of southern Australia

Plant community	Mammals	Birds
Nadgee, N.S.W.		
open-forest	28	59
heathland	19	27
Wyperfield, Vic.		
mallee-broombush	18	60
heathland	16	37

In contrast to the endothermic vertebrates, the species diversity of ectothermic lizards increased

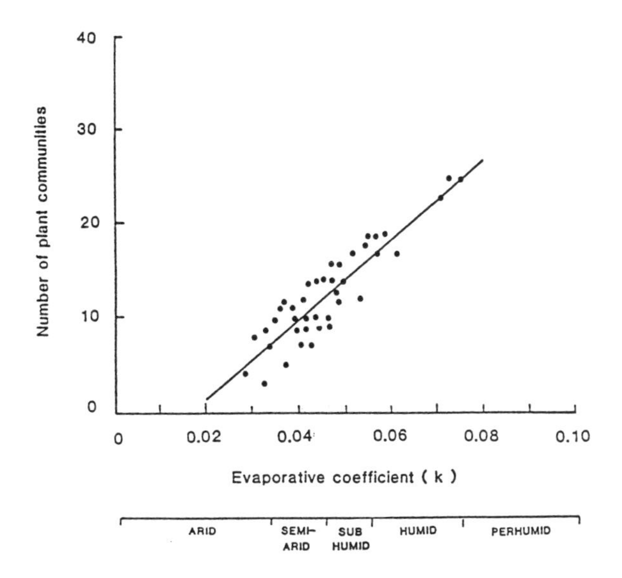

Fig. 2.12. Relationship between the number of major communities (per grid square) in the climatic gradient (evaporative coefficient) from arid to perhumid zones in the mediterranean-climate of South Australia and Victoria (Specht and Specht, 1993).

as more sunlit gaps appear in the vegetation along the climatic gradient from the humid to the arid zone. As species richness of the plant community decreases along this climatic gradient (as annual shoot growth per hectare declines) the number of species of lizards is negatively correlated with species number of vascular plants.

The diversity of species of vascular plants and vertebrates in natural ecosystems in Australia may be predicted for any climatic and edaphic environment. It would appear from the statistics on the diversity of plant species in the many different Mediterranean-type ecosystems throughout the world (Specht, 1988a) and in Israel (Specht *et al.*, 1990) that the paradigms derived from the Australian data are applicable worldwide.

Mediterranean shrublands (for example, chaparral, garrigue, shrubby wetlands, etc.), in which overstorey rootstocks are densely packed, appear to be exceptions to the paradigms which apply to vegetation typical of the Mediterranean-climate regions. After fire, the overstorey regenerates rapidly from the densely packed rootstocks, thus suppressing the growth and diversity of the understorey. In California, some 60 species of vascular plants are found in 0.1 ha of the oak woodland, while the chaparral contains only 24 to 32 species in the same area (Westman, 1988). As the annual shoot growth per hectare of the overstorey of the dense chaparral is only a little lower than the overstorey of the open oak woodland, species diversity of vertebrates is of the same order in the two ecosystems – and apparently higher than the diversity found in the nutrient-poor Australian ecosystems (Catling, 1988). For example, San Dimas Experimental Forest, 7 000 hectares in area, supports 38 mammal, 58 resident bird, and 22 reptile species in the dense Californian chaparral, whereas Wyperfield National Park, 57 000 hectares in area, supports only 18 mammal, 60 resident bird, but more reptile species (46 spp.) in the Australian open-mallee vegetation.

The amount of energy fixed as shoot growth per hectare in the overstorey and understorey strata of plant communities appears to determine the species richness (per hectare) of both vascular plants and endothermic vertebrates. How this process operates is a problem which needs to be investigated, to provide a basis for scientific management of biological diversity in ecosystems.

References

Axelrod, D.I. 1973. History of the Mediterranean ecosystem in California. In: di Castri, F. & Mooney, H.A. (eds), Mediterranean type ecosystems. Springer-Verlag, Berlin, pp. 225–277.

Axelrod, D.I. 1975. Evolution and biogeography of Madrean–Tethyan vegetation. Ann. Miss. Bot. Garden 62: 280–334.

Axelrod, D.I. 1989. Age and origin of chaparral. In: Keeley, S.C. (ed.), The California chaparral. Paradigms reexamined. Natural History Museum, Science Series No. 34, Los Angeles, pp. 7–19.

Braithwaite, R.W., Winter, J.W., Taylor, J.A. & Parker, B.S. 1985. Patterns of diversity and structure of mammalian assemblages in the Australian tropics. Australian Mammalogist 8: 171–186.

Catling, P.C. (co-ordinator) 1988. Vertebrates. In: Specht, R.L. (ed.), Mediterranean-type ecosystems. A Data Source Book. Kluwer Academic Publ., Dordrecht, pp. 171–194.

Edmonds, S.J. & Specht, M.M. 1981. Dark Island heathland, South Australia: faunal rhythms. In: Specht, R.L. (ed.), Ecosystems of the world. Vol. 9B. Heathlands and related shrublands. Analytical studies. Elsevier, Amsterdam, pp. 15–27.

Heddle, E.M. & Specht, R.L. 1975. Dark Island heath (Ninety-Mile Plain, South Australia). VIII. The effect of fertilizers on composition and growth, 1950–1972. Aust. J. Bot. 23: 151–164.

Hill, M.O. 1973. Reciprocal averaging: an eigen vector method of ordination. J. Ecol. 61: 237–249.

Hopkins, B. 1955. The species-area relations of plant communities. J. Ecol. 43: 409–426.

Pons, A. 1981. The history of the Mediterranean shrublands. In: di Castri, F., Goodall, D.W. & Specht, R.L. (eds), Ecosystems of the world. Vol. 11. Mediterranean-type shrublands. Elsevier, Amsterdam, pp. 131–138.

Specht, R.L. 1963. Dark Island heath (Ninety-Mile Plain, South Australia). VII. The effects of fertilisers on composition and growth, 1950–60. Aust. J. Bot. 11: 67–94.

Specht, R.L. 1972. Water use by perennial evergreen plant communities in Australia and Papua New Guinea. Aust. J. Bot. 20: 273–299.

Specht, R.L. 1981. Growth indices – Their role in understanding the growth, structure and distribution of Australian vegetation. Oecologia(Berlin) 50: 347–356.

Specht, R.L. (ed.) 1988a. Mediterranean-type Ecosystems. A Data Source Book. Kluwer Academic Publ., Dordrecht, 248 pp.

Specht, R.L. 1988b. Climatic control of ecomorphological characteristics and species richness in mediterranean ecosystems in Australia. In: Specht, R.L. (ed.), Mediterranean-type ecosystems. A Data Source Book. Kluwer Academic Publ., Dordrecht, pp. 149–155.

Specht, R.L. & Dettmann, M.E. Palaeo-ecology of Australia and current physiological functioning of plant communities. In: Roy, J., Aronson, J. & di Castri, F. (eds), Time scales of biological responses to water constraints. (MEDECOS V, Montpellier) Springer-Verlag, Berlin. (in press)

Specht, R.L. & Morgan, D.G. 1981. The balance between the foliage projective covers of overstorey and understorey strata in Australian vegetation. Aust. J. Ecol. 6: 193–202.

Specht, R.L. & Specht, A. 1989a. Canopy structure in *Eucalyptus*-dominated communities in Australia along climatic gradients. Acta Oecologia, Oecologia Plantarum 10: 191–213.

Specht, R.L. & Specht, A. 1989b. Species richness of overstorey strata in Australian plant communities – the influence of overstorey growth rates. Aust. J. Bot. 37: 321–336.

Specht, R.L. & Specht, A. 1989c. Species richness of sclerophyll (heathy) plant communities in Australia – the influence of overstorey cover. Aust. J. Botany 37: 337–350.

Specht, A. & Specht, R.L. 1993. Species richness and canopy productivity of Australian plant communities. Biodiversity and Conservation 2: 152–167.

Specht, R.L. & Specht, A. 1993. Diversity of plant communities in Australia (in a 1° latitude × 1° longitude grid). Aust. J. Bot. (in press).

Specht, R.L., Grundy, R.I. & Specht, A. 1990. Species richness of plant communities: relationship with community growth and structure. Isr. J. Bot. 39: 465–480.

Specht, R.L., Dettmann, M.E. & Jarzen, D.M. 1992. Community associations and structure in the Late Cretaceous vegetation of southern Australia and Antarctica. Palaeogeog., Palaeoclim., Palaeoecol. 94: 283–309.

Specht, R.L., Bolton, M.P. & Specht, A. 1993. Major plant communities in Australia: – An objective assessment. Aust. J. Bot. (in press).

Westman, W.E. 1988. Vegetation, nutrition and climate: data tables. 3. Species richness. In: Specht, R.L. (ed.), Mediterranean-type Ecosystems. A Data Source Book. Kluwer Academic Publ., Dordrecht, pp. 81–91.

CHAPTER 3

Summergreenness, evergreenness and life history variation in Mediterranean Blue Tits

JACQUES BLONDEL and PAULA CRISTINA DIAS
Centre Louis Emberger, CNRS, B.P. 5051, 34033 Montpellier Cedex, France

Key words: evergreenness, summergreenness, habitat mosaics, caterpillars, Blue Tits, source/sink

Abstract. Using a food chain approach, which includes the leafing patterns of summergreen and evergreen oaks, caterpillars that feed upon them, and insectivorous tits, we examine life history variation of the birds in different habitats of the Mediterranean region. The annual renewal of leaves involves only one third of the whole foliage in the evergreen Holm oak instead of 100% in the summergreen Downy oak and occurs c. three weeks later in the former than in the latter. Caterpillars are nicely synchronized on the leafing process and are more abundant in summergreens than in evergreens. The diet of the Blue Tit includes a large diversity of prey, especially in evergreen oaks, and the way tits compensate for the low abundance of caterpillars is discussed. Blue Tits fairly match the period of food availability in summergreens on the mainland and in isolated habitats dominated by evergreens (i.e. Corsica), but they are mis-timed in evergreen habitat patches within a mainland landscape which includes both summergreen and evergreen habitat patches. We hypothesize that gene flow among sub-populations living in discrete habitat patches of different quality results in a source/sink system whereby the tits breeding in evergeen habitats must be permanently restocked by immigrants from more productive source habitats.

Introduction

One important aspect of plant-animal interactions in Mediterranean-type ecosystems is the ecological and evolutionary consequence of evergreenness (which is associated with sclerophylly) on phytophagous animals and their predators. Investigation of such consequences is particularly relevant in habitat mosaics where the dominant tree species are either summergreen or evergreen. In this chapter, we will examine the consequences of evergreenness versus summergreenness on the breeding and feeding ecology of an insectivorous passerine, the Blue Tit *Parus caeruleus* L. We will first show how and why evergreenness produces changes in the life history of this bird primarily adapted to summergreen habitats of temperate Europe; secondly, we will investigate to what extent birds have evolved local adaptations which match the variation of food availability at the scale of a landscape, including a mosaic of habitat patches dominated either by summergreen or by evergreen trees.

This food chain approach, which considers the relationships between the foliage of the trees, insects which feed upon them and the feeding and breeding biology of the tits, requires collection of data on: 1) the spring development of the vegetation; 2) the seasonal variation of insect abundance; 3) breeding traits of the tits; and 4) the feeding habits of the birds.

Four habitats will be considered in this chapter (Fig. 3.1): 1) a mixed forest on the mainland, at mid-altitude (650–1050 m a.s.l.) along the southern slope of the Mont-Ventoux, southern France. The dominant tree species are the Cedar (*Cedrus atlantica*) and summergreen trees such as the Downy Oak (*Quercus pubescens*) and Maple (*Acer monspessulanus*, *A. opalus*) where the tits mainly forage. This habitat will be hereafter referred as 'Ventoux' or MM (for 'mainland mixed'); 2) an old forest of the evergreen Holm Oak (*Quercus ilex*) at low altitude (100–150 m a.s.l.) near Calvi, on the island of Corsica (IE, for 'island evergreen'); 3) a habitat dominated by the summergreen Downy oak *Quercus pubescens*

M. Arianoutsou and R.H. Groves, Plant-Animal Interactions in Mediterranean-Type Ecosystems, 25–36, 1994.

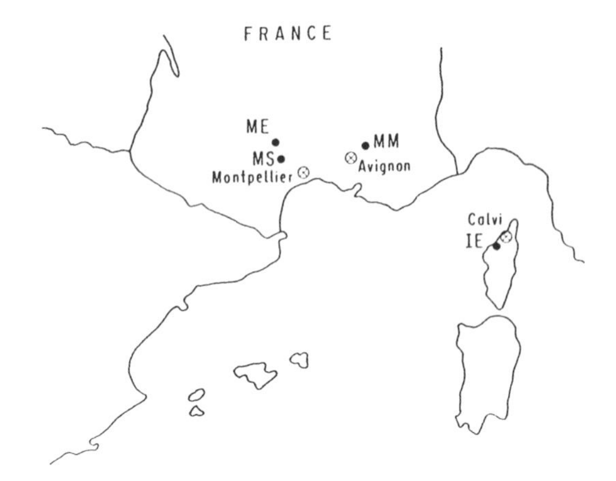

Fig. 3.1. Location of the study sites. MS = mainland summergreen habitat, MM = Mixed habitat (Mont-Ventoux), ME = mainland evergreen habitat, IE = island evergreen habitat (on Corsica).

near Montpellier, southern France (MS, for 'mainland summergreen'); and, 4) a habitat dominated by the evergreen Holm oak *Quercus ilex*, a few kilometres from the latter (ME, for 'mainland evergreen'). The two latter oakwoods are coppices of similar age, c. 40 years, and are parts of the same landscape (see Blondel, 1985; Blondel *et al.*, 1987, 1992, for more details on the habitats and their climate). The leafing process of the trees and the seasonal variation of insect abundance have been studied in the habitats ME, MS and IE; breeding traits of the tits have been collected in the four habitats, and the feeding ecology of tits has been studied in habitats MM (Ventoux) and IE (Corsica). Data were pooled over several years because their between-habitat variation was much higher than their year-to-year within-habitat variation.

The spring development of oaks and caterpilars

Evergreenness involves the yearly renewal of but a fraction of the foliage (*c.* 30%) because leaves are photosynthetically active for several years (Floret *et al.*, 1989). Therefore, since most of the photosynthetic system is present and active all the year round (Eckardt *et al.*, 1975), a large part of energy available in spring for growing processes is allocated to old leaves so that the production of new leaves occurs later and eventually more slowly in evergreen than in summergreen trees, i.e. the spring bloom of the vegetation is less conspicuous in the former than in the latter (Blondel *et al.*, 1992).

The process of spring development of oak leaves has been investigated according to a procedure described by Du Merle and Mazet (1983). We have checked the development of the leaves every 4 days on a scale including 7 stages from stage 1, when the bud is still completely closed, to stage 7, when the new leaf is fully developed. Bud burst (stage 3.5 of the leafing process) provides a good estimation of the spring development of the vegetation and is an important event for tits because it coincides with emergence of the caterpillars on which the birds feed. As expected, bud burst occurred on average 3 weeks later in evergreen oaks both on the mainland and on the island than in summergreen oaks (Table 3.1).

The most important food for tits is caterpillars (Perrins, 1965; Van Balen, 1973; Zandt *et al.*, 1990) which are always the preferred prey whenever available and by far the most important on a biomass basis. Caterpillar abundance was estimated by collecting twice weekly their droppings from the foliage, using 0.25 m^2 trays erected under the canopy of the trees. This method, which has been used for a number of years in tit studies in Holland, reliably estimates the relative abundance of caterpillars and its seasonal variation (Tinbergen, 1960; Zandt *et al.*, 1990). This latter variation roughly parallels that of foliage development: the peak of caterpillar abundance occurred 5 (IE) to 6 (ME) weeks later in evergreen than in summergreen oaks (Table 3.1). Moreover, the total abundance of caterpillars was much lower in evergreen than in summergreen oaks, especially on Corsica (Blondel *et al.*, 1991, 1992). The low abundance of caterpillars in evergreens is presumably related to the fact that leaves older than some months (c. 70% of the whole foliage) are no longer edible by most phyllophagous insects, such as caterpillars, because they are much too hard and contain tannins and other biochemical repellents (Feeny, 1975; Lebreton, 1982). In contrast, all the leaves produced in spring by summergreens are available for leaf-eaters, and hence there is a much higher abun-

Table 3.1. Bud opening of the two species of oaks (stage 3.5 of leaf development), caterpillar peak date, and breeding traits (± 1 SD) of the Blue Tit in four Mediterranean habitats (mean values for 1985–1990 combined)

	Mainland summergreen (MS)	Mainland evergreen (ME)	Mainland mixed (MM)	Island evergreen (IE)
But opening	14 April	6 May		7 May
Caterpillar peak date	30 April	13 June		6 June
Laying date	8 April ± 6.5	18 April ± 6.2	24 April ± 7.5	12 May ± 7.0
Clutch size	9.4 ± 2.0	8.5 ± 1.5	8.6 ± 1.4	6.5 ± 1.1
No. of fledglings	6.4 ± 3.5	3.7 ± 3.4	4.6 ± 3.3	3.2 ± 2.7
Breeding success[1]	0.68	0.43	0.53	0.49

[1]Ratio of the number of fledglings to the number of eggs laid.

dance of insects. Assuming that the food supply proximately and ultimately determines the evolution of such important breeding traits for fitness as laying date and clutch size of birds (Drent & Daan, 1980; Yom-Tov & Hilborn, 1981; Martin, 1987), the large differences in the timing and the abundance of food for tits between evergreens and summergreens predict large differences in the feeding habits and the breeding traits of the birds. Evergreen habitats are supposed to be much poorer as breeding habitats for tits than summergreen ones.

Life history traits of the Blue Tit

The breeding biology of the Blue Tit has been studied using nest-boxes distributed evenly over the habitats at a density of 2 nest-boxes ha^{-1} (100 to 150 nest-boxes depending on habitats and years). Life history traits, especially laying date (mean of the dates on which each pair laid its first egg) and clutch size (mean of the completed clutches laid by each female), have been checked by routine weekly inspections of the nest-boxes over the breeding season. Only first clutches are considered in this chapter because the Blue Tit is usually a single-brood species.

Birds started to breed earlier and laid more eggs in the summergreen than in any evergreen habitat (Table 3.1). The population in the summergreen habitat laid within a few days (6 days on average) around bud opening, which is the time when young caterpillars, especially those of the Green Tortrix *Tortrix viridana*, invade the buds and start to eat the young leaves of oaks (Du Merle & Mazet, 1983) so that they become available as prey for tits. The Corsican population differed from the population in the mainland summergreen habitat by a delay of more than 4 weeks in the onset of breeding and by a 30% reduction of clutch-size (6.5 eggs versus 9.4). The Corsican population also started to lay around the date of local bud opening, however. On average, each breeding pair in the summergreen produced 6.4 fledglings compared with 3.2 in the island evergreen habitat (Table 3.1). The breeding success was higher (number of fledglings per egg = 0.68) in the former than in the latter (number of fledglings per egg = 0.49). On the other hand, in the mainland evergreen habitat, which closely matches the Corsican evergreen as concerns the seasonal variation of food, Blue Tits started to lay nearly 3 weeks before local bud opening and laid more eggs than expected on the basis of the Corsican figures. Actually, breeding traits in this habitat (ME) are more similar to those in the nearby summergreen habitat (MS). As a result, the breeding success (number of fledglings per egg = 0.43) was especially low in ME, presumably because tits laid too early in relation to food availability so that they mismatched the best period of caterpillar abundance. This mistiming in the evergreen mainland habitat will be discussed later (see subsequent section Summergreenness and evergreenness in habitat mosaics). Finally, life history traits in the mainland mixed habitat had somewhat intermediate values between those in the summergreen and those in the evergreen oakwoods, mostly because of altitude effects (see Blondel, 1985).

Food and feeding habits of the Blue Tit

The feeding habits of tits were studied in the mainland mixed forest (MM, Ventoux) and in the evergreen island habitat (IE) through the food items brought to the nestlings by their parents. For this purpose, a nest-box was transformed into an automatic camera system (see Blondel *et al.*, 1991 for details). Field work included 6 nests (totaling 20 full days filming and 1,662 successful pictures) on Ventoux (MM) and 7 nests (35 days, 2,138 pictures) on Corsica (IE). For each photograph, the following parameters were recorded: time of entry, sex of the parent, type and life-form of prey, length and width of prey. Prey items were identified as far as possible in the taxonomic hierarchy (see Blondel *et al.*, 1991).

The data set included 69 taxa on Corsica and 54 on Ventoux. They have been lumped into eight broad categories in Fig. 3.2, i.e. Lepidoptera (mostly larvae), Spiders, Coleoptera, Diptera, Orthoptera, Hemiptera, Dermaptera and 'other insects' (mainly Hymenoptera). Caterpillars constituted the major part of the diet on Ventoux (57.5%), but only 17.4% on Corsica, which is a surprisingly low figure because in all studies so far carried out on the diet of tits, caterpillars have always been the dominant prey category (Blondel *et al.*, 1991). Spiders were an important food in both habitats, especially on Corsica where they were the dominant prey category. The third most important prey category on Corsica were grasshoppers (17.3%). The remaining prey items belonged to a wide range of families including Coleoptera, Diptera, Hemiptera and Dermaptera. As a consequence of the poor contribution of caterpillars, the diet was much more diverse on the island than on the mainland as illustrated by differences in the diversity indices H′ (Shannon's function) which were always higher on Corsica, not only at the level of the whole diet, but also at that of each prey category (Fig. 3.2).

These differences in the taxonomic composition of the diet are illustrated in Fig. 3.3 where the proportion of caterpillars is plotted against that of other prey categories for full-day filming sequences in the two sites. On Ventoux, the main alternative prey items to caterpillars were spiders. On Corsica, the consistently very low proportion of caterpillars was compensated by several other

(a)

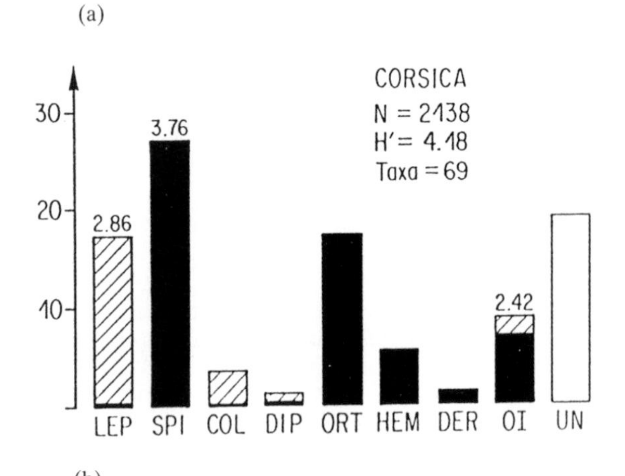

(b)

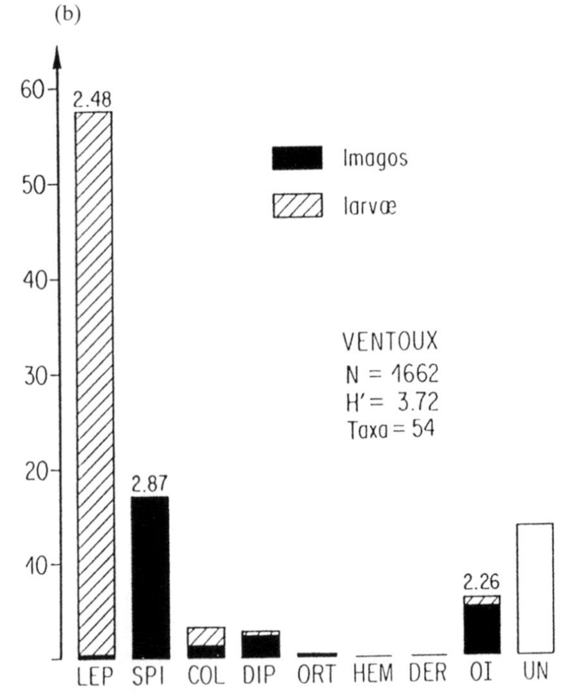

Fig. 3.2. Composition of the diet (in percent) of Blue Tit nestlings on Corsica (upper) and Ventoux (lower). LEP = Lepidoptera, SPI = Spiders, COL = Coleoptera, DIP = Diptera, ORT = Orthoptera, HEM = Hemiptera, DER = Dermaptera, OI = Other insects, UN = unidentified. N = sample size. H′ = Diversity index (Shannon). Numbers above the columns are indices of diversity within prey categories (Blondel *et al.*, 1991).

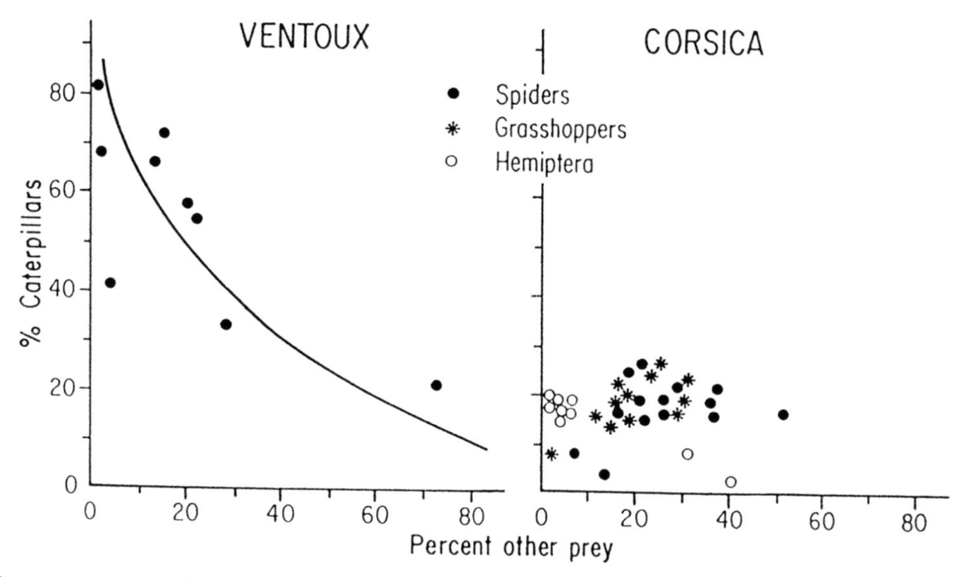

Fig. 3.3. Percentage of caterpillars in the diet of nestling Blue Tits in relation to the percentage of other prey types. The line for Ventoux has been fitted by eye from 9 full-day filming sequences; there were 14 such sequences on Corsica. See text for further explanation.

categories, mainly spiders, grasshoppers and Hemiptera. Caterpillars, which are the most profitable prey, were much larger on Corsica than on Ventoux, however. In all cases, the standard deviation of prey size, either length or width, was much larger on the island: taxonomic diversity on the island was associated with a large diversity of size and shape. All the pairs photographed showed the same pattern.

Because a volume index gives a better picture of the amount of food brought to the young than prey size, such an index has been calculated for the two main prey categories, caterpillars and spiders (see Blondel *et al.*, 1991). These volume indices strongly increase the inter-habitat differences of prey size. Caterpillars were on the average much larger on Corsica than on the mainland (+60.3%). Average feeding frequencies by the parent tits were significantly lower on Corsica (3.90 ± 0.95 feeds per chick per hour) than on Ventoux (5.41 ± 2.8, $\chi^2(1) = 13.16$, $P = 0.00029$).

In summary, by comparison with the mainland, food and feeding patterns of Corsican Blue Tits include a smaller proportion of caterpillars, a larger diversity of prey items, larger prey, and lower feeding frequencies.

How do tits compensate for the low caterpillar abundance in evergeen habitats?

Three sets of factors must allow the tits to meet their food requirements, especially in the poor evergreen environment of Corsica:

1. Evergreen Mediterranean ecosystems are characterized by a large diversity of arthropods, few of them abundant as a species (Lack, 1966; Owen, 1977; Du Merle, pers. comm.). A relative environmental stability in the form of the seasonal constancy of the climate in Mediterranean-type ecosystems probably enhances arthropod diversity as compared with ecosystems in central Europe which are characterized by a much stronger seasonality. Therefore, a late production of a diverse albeit population-poor set of arthropods in evergreen habitats contrasts with a short but highly productive bloom of a few abundant species such as the Green Tortrix or the Winter Moth *Operophtera brumata* in summergreen habitats. Since varying the diet is probably much less advantageous than collecting as much as possible of the same type of food (Royama, 1970; Owen, 1977; Rabenold, 1978) a diversified diet in Mediterranean ever-

Table 3.2. Comparison of the diet of the Blue Tit in Mediterranean habitats and various habitats in Europe. Data expressed as percent of the total diet. Unidentified preys excluded. LEP = Lepidoptera, SPI = Spiders, ORT = Orthoptera, DIP = Diptera, HYM = Hymenoptera, HEM = Hemiptera, COL = Coleoptera. Diversity calculated from the Shannon's function H' (log2)

Habitat	LEP			SPI	ORT	DIP	HYM	HEM	COL	Diversity (H')	Reference
	Larvae	Pupae	Imago								
1 Oakwood	84.4	0.3	0.0	9.7	0.0	1.7	0.0	0.0	0.0	0.47	Minot (1981)
2 Oakwood[1]	19.0	72.0	0.0	6.3	0.0	0.0	0.0	0.0	0.0	0.74	Betts (1955)
3 Oakwood	65.0	0.7	0.0	21.9	0.0	2.3	0.0	6.8	3.0	1.46	Török (1986)
4 Scots Pines	46.8	14.2	0.8	21.2	0.0	0.2	0.2	0.2	0.2	1.86	Gibb and Betts (1963)
5 Mixed (Ventoux)	57.7	1.0	5.7	19.3	0.3	2.8	0.5	0.0	1.1	1.89	This study
6 Gardens (Wales)[2]	42.6	1.7	0.5	4.8	0.0	0.2	3.7	6.2	0.2	2.01	Cowie and Hinsley (1988)
7 Corsican Pines	43.0	2.5	10.5	24.5	0.0	2.0	1.0	9.5	2.0	2.16	Gibb and Betts (1963)
8 Holm oak (Corsica)	17.2	0.0	0.2	33.2	21.8	1.5	0.1	8.2	0.1	2.19	This study

[1]Data on this habitat were collected at one nest late in the season (Betts, 1955), which explains the high proportion of Lepidoptera pupae.
[2]This data set includes 14.9% artificial food which has been included in the calculation of diversity.

green habitats reflects the absence of any superabundant food supply. This is consistent with foraging theory predictions according to which diversity of prey taken by a predator increases with decreased abundance of food (Krebs, 1973; Rabenold, 1978). In optimal habitats of temperate Europe the Blue Tit forages in oak leaves where it mainly searches for superabundant caterpillars (Gibb, 1954, 1960; Snow, 1954; Tinbergen, 1960; Partridge, 1974). In Mediterranean habitats, this bird has frequently been observed to forage in a wide range of microsites from the ground up to the canopy of the trees (pers. obs.), thereby taking advantage of the wide range of foraging techniques and flexibility in food collecting which characterize this species (Gibb, 1954, 1960; Partridge, 1976). As a rule, food diversification and switching to prey other than caterpillars occurs in suboptimal habitats dominated by conifers: mixed stands, Scots pines, Corsican pines. For this reason, the diversity of the diet is lowest in optimal oakwood where most of the food supply is made up of the same food category (caterpillars) and then increases as the habitat becomes less optimal. Eight habitats for which data are available can be classified into three categories (Table 3.2): 1) optimal summergreen oakwood where the diversity of the diet H' is the lowest (habitats 1,2,3); 2) mixed and coniferous woods (and gardens) where H' has intermediate values (habitats 4,5,6,7); and 3) the Corsican evergreen Holm Oak where caterpillar abundance is especially low (habitat 8). Lepidoptera at all stages of development constitute on average 80.5% of the diet in summergreen oakwood and then decrease to 60.7% in mixed stands and conifers, 44.8% in gardens, and 17.4% in Corsican evergreen oakwood.

2. Larger food items, especially caterpillars, partly compensate for the poor caterpillar abundance on Corsica. The reason why tits do not take young small caterpillars earlier in the season is unknown, however.
3. Ambient temperature is on average 7.5 °C higher during the nestling period on Corsica (20.1 °C, years 1976–88 combined) than on Ventoux (12.6 °C). It is often near the physiological optimum during daytime so that energetic requirements are lower on the island. The late nestling period in the warm Corsican environment must be an important saving of energy because food consumption decreases as air temperature increases (Royama, 1966; Mertens, 1969). For instance, Dutch Great Tits decrease food consumption at daytime temperatures over 23 °C (Van Balen, 1973). Because daytime temperature is generally higher than 23 °C on Corsica during the nestling period, food requirements are expected to be lower there as compared with those in temperate habitats (ambient temperature was c. 15 °C in the Van Balen's study). On the other hand, an increased effort in evaporative cooling by panting during very warm periods

at the end of the nestling period (temperature may exceed 30 °C in the nest-boxes) may be energy consuming (Calder & King, 1974; Yom-Tov & Hilborn, 1981; Mertens, 1988).

Food abundance and life history traits

The quality and the quantity of the food supply have been recognized for long as major determinants in shaping breeding traits of birds (Perrins, 1970; Ankney & MacInnes, 1978; Drent & Daan, 1980; Yom-Tov & Hilborn, 1981; Martin 1987; Arcese & Smith, 1988). Accordingly, insectivorous birds should be timed in such a way that the peak of arthropod production coincides with the nestling period, a time in the reproductive season that makes the greatest demand on the adults' food gathering capacity (Lack, 1966; Rabenold, 1978; Martin, 1987).

Studies carried out up to now in summergreen oakwoods of non-Mediterranean Europe (Hartley, 1953; Betts, 1955; Gibb, 1950, 1955, 1960; Royama, 1966, 1970; Van Balen, 1973; Minot, 1981; Török, 1986; Cowie & Hinsley, 1988) have shown that caterpillars are usually very abundant in summergreen oak woodland and constitute the bulk of the food of the tits. Studies in which the variation of caterpillar availability has been studied in relation to the breeding biology of tits, have shown a positive correlation between the mean laying date and the caterpillar peak date (Gibb, 1950; Van Balen, 1973; Perrins & McCleery, 1989; Zandt *et al.*, 1990, Fig. 3.4). The relationship between laying date and the peak of caterpillar abundance also fits this correlation in the mainland summergreen and the Corsican evergreen habitats (Fig. 3.4). This relationship suggests that caterpillars are also a key prey for tits in Mediterranean habitats, including on Corsica, in spite of their very low abundance. Such a correlation indicates that the timing of the breeding season has evolved in such a way that maximal food demand by the young (when they are about ten days old) roughly coincides with the peak of caterpillar abundance (Perrins, 1965; Van Balen, 1973; Zandt *et al.*, 1990). The very late laying date in Corsica is assumed to be a response to the late development of evergreen trees and to that of their associated arthropod

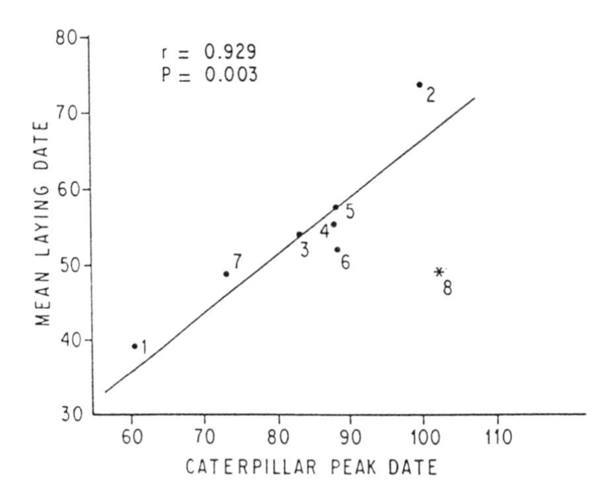

Fig. 3.4. Correlation between laying date of the tits (Great Tit or Blue Tit) and the peak date of caterpillars (dates reckoned from 1 March, i.e. 1 = 1st March, 32 = 1st April) in several European oakwoods. 1 = summergreen mainland Mediterranean habitat (MS), 2 = evergreen island Mediterranean habitat (IE, Corsica), 3 = Holland, 4 = Great Britain, 5 = Holland, 6 = Holland, 7 = Switzerland, 8 = evergreen mainland Mediterranean habitat (ME). Because tits are mistimed in this latter habitat (gene flow from richer nearby habitats, see text) it has not been included in calculating the correlation (see text; completed after Zandt *et al.*, 1990).

fauna. However, as stated above, laying date in the mainland evergreen habitat (habitat 8 on Fig. 3.4) does not fit the trend. This will be discussed subsequently.

Clutch size on Corsica is much lower than expected if reduction of clutch size were only due to such factors as age of the female, breeding density, date of laying or territory quality, which are all known to have negative effects on clutch-size (Blondel, 1985). Clearly, the most important factor is food which is widely recognized as the ultimate factor in the evolution of both laying date and clutch size (Lack, 1950; Perrins, 1965, 1970; Immelmann, 1971; Martin, 1987; Perrins & McCleery, 1989). Most of the variation in clutch-size is usually interpreted as an adaptation to the food supply for egg formation and for rearing a number of young which in turn is limited by the amount of food the parents can collect (Klomp, 1970; Van Balen, 1973; Martin, 1987; Nur, 1988; Pettifor *et al.*, 1988; Slagsvold & Lifjeld, 1988). Thus the low clutch size on Corsica is hypothesized to be ultimately an adaptation to poor food resources.

Because parents must compromise between clutch quality (hence probability of survival of their young) and future reproductive prospects (hence probability of their own survival), fecundity is set by the additional mortality of both young and adults that is associated with the breeding effort, i.e. by reproductive mortality (Williams, 1966; Charnov & Krebs, 1974). Compared with the mainland summergreen habitat, the late onset of breeding, smaller clutch size, large diversity of the diet, and the reduction in food abundance are all consistent with the hypothesis that food limitation is the underlying cause of the pattern (see review by Martin, 1987). These results support to some extent Lack's food limitation hypothesis (Lack, 1968), according to which clutch-size of passerines is reduced on Mediterranean islands because of scarcity of food. However, the scarcity of food seems to be less a feature of insularity than a property of Mediterranean evergreenness. Other factors that could have played an important role in the evolution of clutch size on Corsica are the combination of a water-deficient food supply, which may cause problems of water balance for the nestlings, and high temperature which may lead to hyperthermia. The percentage of heat produced by nestlings which can be dissipated by evaporation of body water strongly depends on the water content of prey (Mertens, 1977). Water content largely differs according to prey types: Around 85% for caterpillars, 73% for spiders and 70% for grasshoppers (Edney, 1977). On Corsica, the combination of high temperature with a large proportion of water-deficient prey items, such as spiders and grasshoppers, could lead to problems for the water balance of young in the dry-warm nest-box environment and thereby may limit reproduction, and hence clutch-size (Blondel *et al.*, 1991).

To sum up, tits lay as early as they are able to do in relation to the food supply, and produce a clutch adapted to the number of young they can raise. In the rich summergreen habitat on the mainland, tits lay many eggs and start to breed early, because food is plentiful and is available early, whereas in the poor evergreen habitat on Corsica, clutch size is much lower and the laying date occurs later because food is scarcer and is available even later in the season. Thus the large between-habitat variation of breeding traits of insectivorous birds such as tits in Mediterranean habitats is related to whether the dominant tree species, where the birds forage, are evergreen or summergreen.

Summergreenness and evergreenness in habitat mosaics

One puzzling feature of breeding traits of the Blue Tit in the systems under study is the intermediate values of both clutch-size and laying date in the mainland evergreen habitat (ME) as compared to the mainland summergreen (MS) and the island evergreen (IE) habitats. Since the leafing process of oaks and the variation of caterpillar abundance were similar in both the mainland and the island evergreen habitats, life history traits of the Blue Tit were also expected to be of a similar order of magnitude. This is not the case, however, because tits in the poor evergreen habitat on the mainland started to breed too early and laid too many eggs in relation to the seasonal variation of the food supply (Fig. 3.5). As a result, many nestlings died from starvation so that the breeding success was very low (0.43, Table 3.1, Fig. 3.5). Why are breeding traits in this habitat more similar to those in the rich neighbouring summergreen habitat and why do they not follow the same trends as those on Corsica?

One hypothesis to explain this mistiming and mismatching of tits to the patterns of food availability is the gene flow hypothesis among subpopulations living in discrete habitat patches of different quality within a landscape (Blondel *et al.*, 1992). The mosaic of rich and poor habitat patches is hypothesized to operate as a source/sink system (Pulliam, 1988). Poor sink habitats, where birds are misadapted because they immigrate from rich source-habitats to which they are nicely adjusted, produce few recruits so that reproduction cannot balance local mortality. Populations in such habitats can persist provided that they are permanently restocked by immigrants from more productive source-habitats. Such a hypothesis assumes that both laying date and clutch size have a strong genetic basis. Heritability of

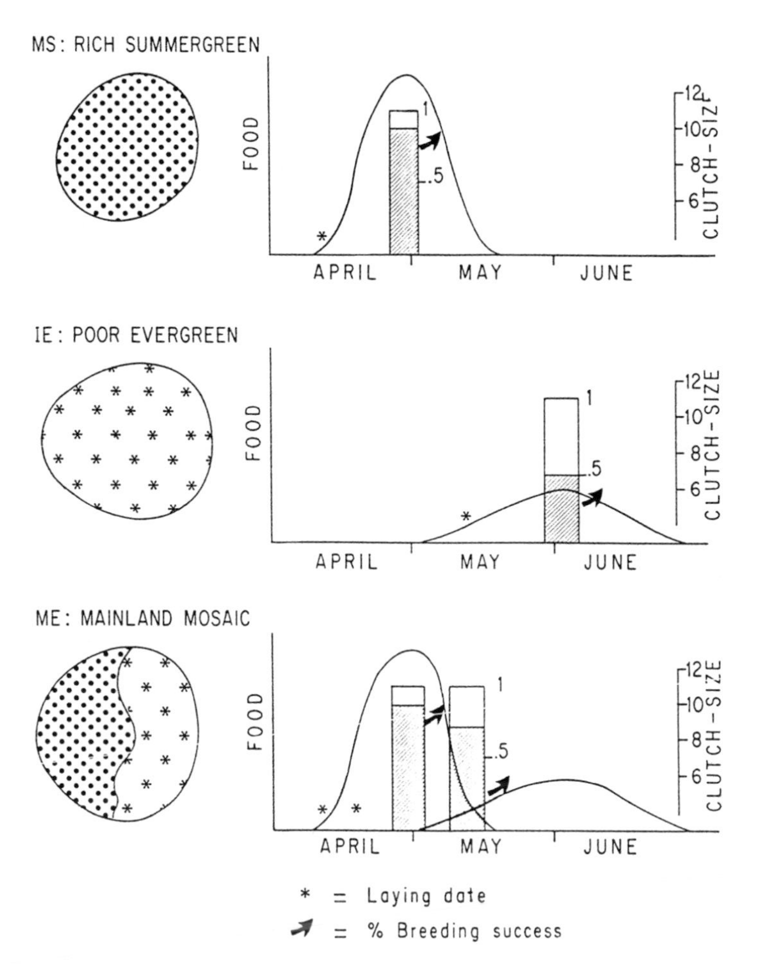

Fig. 3.5. Relationship between breeding traits of the tits and the seasonal variation of the food supply. Columns indicate clutch size and their location on the horizontal axis shows to what extent the breeding time matches the peak of caterpillar abundance. There is a good match for populations breeding in both the mainland summergreen (MS) and the island evergreen (IE) habitats but obvious mis-matching for the population breeding in the mainland evergreen (ME) habitat, presumably because of gene flow from the nearby rich MS habitat. Stars indicate laying date. For a clutch-size of 10 eggs, birds match fairly well the food resources if they lay on average 34 days before the peakdate of caterpillars (10 days for completing the clutch because birds lay 1 egg per day + 14 days of incubation + 10 days, because the maximum demand of food by the young occurs when they are about 10 days old). Arrows indicate breeding success (= number of fledglings/number of eggs laid, total height of the column = 100% breeding success).

breeding traits has been demonstrated by Van Noordwijk *et al.* (1980) for the Great Tit *Parus major* and for Mediterranean populations of Blue Tits by Blondel *et al.* (1990). However, the slightly smaller values of breeding traits in the evergreen mainland oakwood than in the summergreen one, from which birds are supposed to immigrate, show that some adjustment of traits to local conditions may be achieved through phenotypic plasticity. Concerning the mechanisms of asymmetric dispersal between source and sink habitats, what makes a bird decide to stay in or to move from the habitat where it was born? Differences in social status are known to affect dispersal and the distribution of individuals (Fretwell, 1972). Since survival and dispersal are partly determined by the rank of individuals in a social hierarchy (De Laet, 1985), the effect of the habitat, with Blue Tits reared in rich summergreen oakwoods dominating individuals reared in poor evergreen ones (Dias *et al.*, in press), must affect the habitat distribution of individuals (Westman, 1990). It is hypothesized that individuals of low dominance rank will be eliminated from better summergreen habitats and settle in the poor evergreen ones, thereby leading to a despotic distribution (Fretwell & Lucas, 1970). The results of studies in progress suggest that such is the case. However, because the dispersal range of individuals is larger than the size of each habitat patch, and because some dominant birds can settle in poor habitats and some inferior birds can settle in rich habitats, it is unlikely that asymmetric dispersal rates result in a marked segregation among Blue Tit populations in Mediterranean landscapes. On the other hand, since the Corsican Blue Tit has been isolated from mainland populations for a long time and has evolved as a subspecies (*P. c. ogliastrae*) in habitats where there is no more rich summergreen habitat within the range of dispersal of the population, life history traits of the bird have evolved in such a way that they fairly well match local environmental conditions.

Concluding remarks

Two points merit attention. First, concerning issues on island biology, a more diversified diet on the island than on the mainland supports the hypothesis of niche broadening in insular populations, a feature described by many authors as a response to a release in interspecific competition in highly depauperate island communities (MacArthur *et al.*, 1972; Williamson, 1981; Blondel, 1986). Indeed, the total species richness of birds on Corsica is reduced by *c.* 30% compared with the richness in areas of similar size on the nearby mainland (Blondel *et al.*, 1988). However, although Corsican biotas exhibit several features of a so-called 'insular syndrome', such as the species impoverishment just mentioned, the mechanisms of such a broadening of the food niche of insectivorous birds have probably much less to do with changes in competition at the level of community than with local constraints related to the poor abundance of insect populations in evergreen Mediterranean habitats. This case study is a counter example to the broad generalization on the mechanisms of niche broadening on islands.

Second, the differences in leafing habits between summergreen and evergreen oaks have consequences at both the level of populations and that of communities of insectivorous birds. At the population level, living in habitats of different quality may result in such changes in the breeding season, fecundity, survival, and dispersal patterns that trade-offs between traits are completely different depending on habitat quality. This may be an important issue in species conservation because population studies in poor (sink) habitats may yield only weak information on the factors regulating population size, if population size in such habitats largely depends on the source populations of neighbouring rich habitats. In landscapes that include different habitat patches, the contribution of each patch to the total population and the evolution of life history traits depend on the relative size of habitat patches. In turn, such changes may have consequences at the community level. Most bird species are more sensitive to the structure of the vegetation than to the floristic composition (Cody, 1981), hence they can be found in a variety of habitats. If the dispersal range of a species is larger than the size of habitat patches, then those individuals that settle in poor habitats may become maladapted to local conditions, produce very few offspring and eventually become genetically dead. The persistence of populations in patchy environments, hence some components of community diversity, may depend on habitat-specific demographic rates in other habitats. Thus, an increased understanding of community structure needs to consider neighbouring habitats because the diversity in any given habitat may depend on both the regional diversity of habitats and the diversity and abundance of resources in each local habitat. Modelling the evolution of genetic and demographic systems in such situations may be useful for landscape management decisions related to conservation biology.

Acknowledgments

We are very grateful to Richard Groves who gave

us useful suggestions and improved an early draft of this chapter.

References

Ankney, C.D. & MacInnes, C.D. 1978. Nutrient reserves and reproductive performance of female Lesser Snow Geese. Auk 95: 459–471.

Arcese, P. & Smith. J.N.M. 1988. Effects of population density and supplemental food on reproduction in song sparrows. J. Anim. Ecol. 57: 119–136.

Betts, M.M. 1955. The food of titmice in oak woodlands. J. Anim. Ecol. 24: 282–323.

Blondel, J. 1985. Breeding strategies of the Blue Tit and the Coal Tit (*Parus*) in mainland and island Mediterranean habitats: a comparison. J. Anim. Ecol. 54: 531–556.

Blondel, J. 1986. Biogéographie Evolutive. Masson, Paris.

Blondel, J., Chessel, D. & Frochot, B. 1988. Bird species impoverishment, niche expansion and density inflation in Mediterranean island habitats. Ecology 69: 1899–1917.

Blondel J., Clamens, A., Cramm, P., Gaubert, H. & Isenmann, P. 1987. Population studies of tits in the Mediterranean region. Ardea 75: 21–34.

Blondel, J., Dervieux, A., Maistre, M. & Perret, Ph. 1991. Feeding ecology and life history variation of the Blue Tit in Mediterranean mainland and island habitats. Oecologia 88: 9–14.

Blondel, J., Perret, Ph., Maistre, M. & Dias, P. 1992. Do harlequin Mediterranean environments function as source sink for Blue Tits (*Parus caeruleus* L.)? Landscape Ecology 6: 213–219.

Blondel, J., Perret, Ph. & Maistre, M. 1990. On the genetical basis of the laying date in an island population of Blue Tit. J. Evol. Biol. 3: 469–475.

Calder, W.A. & King, J.R. 1974. Thermal and caloric relations of birds. In Farner, D.S. & King, J.R. (eds.). Avian Biology, vol. IV. Academic Press, New York and London, pp. 259–413.

Charnov, E.L. & Krebs, J.R., 1974. On clutch-size and fitness. Ibis 116: 217–219.

Cody, M.L. 1981. Habitat selection in birds: the roles of vegetation structure, competitors, and productivity. BioScience 31: 107–113.

Cowie, R.J. & Hinsley, S.A. 1988. Feeding ecology of Great Tits (*Parus major*) and Blue Tits (*Parus caeruleus*), breeding in suburban gardens. J. Anim. Ecol. 57: 611–626.

De Laet, J.V. 1985. Dominance and aggression in juvenile great tits, *Parus major* L. in relation to dispersal. In: Sibly, R.M. & Smith, R.H. (eds.). Behavioural Ecology. Ecological Consequences of Adaptive Behaviour. Blackwell, Oxford, pp. 375–380.

Dias, P. C., Meunier, F., Beltra, S. & Cartan-Son, M. 1993. Blue Tits in Mediterranean Mosaics. Ardea (in press).

Drent, R.H. & Daan, S. 1980. The prudent parent: energetic adjustments in avian breeding. Ardea 68: 225–252.

Du Merle, P. & Mazet, R. 1983. Stades phénologiques et infestation par *Tortrix viridana* L. (Lep., Tortricidae) des bourgeons du chêne pubescent et du chêne vert. Acta Oecologica/Oecol. Applic. 4: 47–53.

Eckardt, F.E., Heim, G., Methy, M. & Sauvezon, R. 1975. Interception de l'énergie rayonnante, échanges gazeux et croissance dans une forêt méditerranéenne à feuillage persistant (*Quercetum ilicis*). Photosynthetica: 145–156.

Edney, E.B. 1977. Water Balance in Land Arthropods. Springer Verlag, Berlin, Heidelberg, New York.

Feeny, P.P. 1975. Biochemical coevolution between plants and their insect herbivores. In: Gilbert, L.E. & Raven, P.H. (eds.). Coevolution of Plants and Animals. Univ. Texas Press, Texas, pp. 3–19.

Floret, Ch., Galan, M.J., Le Floc'h, E., Leprince, F. & Romane, F. 1989. In: Orshan, G. (ed.). Plant Pheno-morphological Studies in Mediterranean Type Ecosystems. Kluwer Academic Publ., Dordrecht, pp.9–97.

Fretwell, S.D. 1972. Populations in a Seasonal Environment. Princeton Univ. Press, Princeton, N.J.

Fretwell, S.D. & Lucas, H.L. 1970. On territorial behaviour and other factors influencing habitat distribution in birds. Acta Biotheor. 19: 16–36.

Gibb, J.A., 1950. The breeding biology of the Great and Blue Titmice. Ibis 92: 507–539.

Gibb, J.A. 1954. Feeding ecology of tits, with notes on Treecreeper and Goldcrest. Ibis 96: 513–543.

Gibb, J.A. 1955. Feeding rates of Great Tits. Brit. Birds 48: 49–58.

Gibb, J.A., 1960. Populations of tits and goldcrests and their food supply in pine plantations. Ibis 102: 163–208.

Gibb, J.A. & Betts, M.M. 1963. Food and food supply of nestling tits in Breckland pine. J. Anim. Ecol. 32: 489–533.

Hartley, P.H.T. 1953. An ecological study of the feeding habits of the English titmice. J. Anim. Ecol. 22: 261–288.

Immelmann, K. 1971. Ecological aspects of periodic reproduction. In: Farner, D.S. & King, J.R. (eds.). Avian Biology, vol. 1. Academic Press, New York and London, pp. 341–389.

Klomp, H. 1970. The determination of clutch size in birds. A review. Ardea 58: 1–124.

Kluyver, H.N. 1950. Daily routines of the Great Tit, *Parus m. major* L. Ardea 38: 99–135.

Krebs, J.R. 1973. Behavioural aspects of predation. In: Bateson, P.P.G. & Klopfer, P.H. (eds.). Perspectives in Ethology. Plenum Press, New York, pp. 73–111.

Lack, D. 1950. The breeding seasons of European birds. Ibis 92: 288–316.

Lack, D. 1966. Population Studies of Birds. Clarendon Press, Oxford, England.

Lack, D. 1968. Ecological Adaptations for Breeding in Birds. Methuen, London.

Lebreton, Ph. 1982. Tanins ou alcaloïdes: deux tactiques phytochimiques de dissuasion des herbivores. Rev. Ecol. (Terre et Vie) 36: 539–572.

MacArthur, R.H., Karr, J.R. & Diamond, J.M. 1972. Density compensation in island faunas. Ecology 53: 330–342.

Martin, T.E. 1987. Food as a limit on breeding birds: a life-history perspective. Annu. Rev. Ecol. Syst. 18: 453–487.

Mertens, J.A.L., 1969. The influence of brood size on the

energy metabolism and water loss of nestling Great Tits *Parus major*. Ibis 111: 11–16.

Mertens, J.A.L. 1977. Thermal conditions for successful breeding in Great Tits (*Parus major* L.), I. Relation of growth and development of temperature regulation in nesting Great Tits. Oecologia 28: 1–29.

Mertens, J.A.L. 1988. Water balance of Blue Tits (*Parus caeruleus*) in Corsica. Institute for Ecological Research, Progress Report 1987: 13–18.

Minot, E.O. 1981. Effects of interspecific competition for food in breeding blue tits (*Parus caeruleus*) and great tits (*Parus major*). J. Anim. Ecol. 50: 375–386.

Nur, N. 1988. The cost of reproduction in birds: An examination of the evidence. Ardea 76: 155–168.

Owen, D.F. 1977. Latitudinal gradients of clutch size: an extension of David Lack's theory. In: Stonehouse, B. & Perrins, C.M. (eds.). Evolutionary Ecology. Macmillan, London, pp. 170–180.

Partridge, L. 1974. Habitat selection in Titmice. Nature, London 247: 573–574.

Partridge, L. 1976. Field and laboratory observations on the foraging and feeding techniques of Blue Tits (*Parus caeruleus*) and Coal Tits (*Parus ater*) in relation to their habitats. Anim. Behav. 24: 534–544.

Perrins, C.M. 1965. Population fluctuations and clutch-size in the Great tit (*Parus major*). J. Anim. Ecol. 34: 601–647.

Perrins, C.M. 1970. The timing of birds' breeding seasons. Ibis 112: 242–255.

Perrins, C.M. & McCleery, R.H. 1989. Laying dates and clutch size in the Great Tit. Wilson Bull. 1091: 236–253.

Pettifor, R.A., Perrins, C.M. & McCleery, R.H. 1988. Individual optimization of clutch size in great tits. Nature 336: 160–162.

Pulliam, H.R. 1988. Sources, sinks, and population regulation. Am. Nat. 132: 652–661.

Rabenold, K.N. 1978. Foraging strategies, diversity, and seasonality in bird communities of Appalachian Spruce-Fir forests. Ecol. Monogr. 48: 397–424.

Royama, T. 1966. Factors governing feeding rate, food requirement and brood-size of nestling Great Tits *Parus major*. Ibis 108: 313–347.

Royama, T. 1970. Factors governing the hunting behaviour and selection of food by the Great tit (*Parus major* L.). J. Anim. Ecol. 39: 619–668.

Slagsvold, T. & Lifjeld, J.T. 1988. Ultimate adjustment of clutch size to parental feeding capacity in a passerine bird. Ecology 69: 1918–1922.

Snow, D.W., 1954. The habitats of Eurasian tits, *Parus* spp. Ibis 96: 565–585.

Tinbergen, L. 1960. The natural control of insects in pinewoods. I. Factors influencing the intensity of predation by songbirds. Archives Néerlandaises de Zoologie 13: 265–336.

Török, J. 1986. Food segregation in three hole-nesting bird species during the breeding season. Ardea 74: 129–136.

Van Balen, J.H. 1973. A comparative study of the breeding ecology of the Great Tit *Parus major* in different habitats. Ardea 61: 1–93.

Van Noordwijk, A.J., Van Balen, J.H. & Scharloo, W. 1980. Heritability of ecologically important traits in the great tit. Ardea 68: 193–203.

Westman, B. 1990. Environmental effect on dominance in young Great Tits *Parus major*: a cross-fostering experiment. Ornis Scand. 21: 46–51.

Williams, G.C. 1966. Natural selection, the cost of reproduction, and a refinement of Lack's principle. Am. Nat. 100: 687–690.

Williamson, M. 1981. Island Populations. Oxford Univ. Press.

Yom-Tov, Y. and Hilborn, R. 1981. Energetic constraints on clutch size and time of breeding in temperate zone birds. Oecologia 48: 234–243.

Zandt, H., Strijkstra, A., Blondel, J. & Van Balen, H. 1990. Food in two Mediterranean Blue Tit populations: Do differences in caterpillar availability explain differences in timing of the breeding season? In: Blondel, J., Gosler, A., Lebreton, J.D. & McCleery, R. (eds.). Population Biology of Passerine Birds, An Integrated Approach. NATO ASI Series G, vol 24. Springer-Verlag, Berlin, Heidelberg, pp. 145–155.

CHAPTER 4

Community structure and species richness in the Mediterranean-type soil fauna

A. LEGAKIS
Zoological Museum, Department of Biology, University of Athens, Greece

Key words: Mediterranean-type ecosystems, soil fauna, community structure, species richness

Abstract. Recent results on community structure and species richness of the soil fauna of Mediterranean-type ecosystems are reviewed. Faunal composition, relative abundance of animal groups and trophic levels, spatial and temporal variation of the community structure together with species diversity are summarized. The effects of fire, grazing, agriculture and of other human activities on community structure are described. It is concluded that the soil fauna system of the Mediterranean-type ecosystems has a significant spatial, temporal and ecological complexity that can be better understood only by detailed species-level, long-term, seasonal and multi-sample studies.

Introduction

Ecological aspects of the soil fauna of Mediterranean-type ecosystems are poorly known. Only three large-scale reviews have appeared over the last 20 years (Di Castri, 1973; Di Castri & Vitali-Di Castri, 1981; Majer & Greenslade, 1988). Most of the specific articles that have appeared deal with the ecology of particular animal groups or species whilst few are concerned with the community as a whole. This chapter will attempt to summarize some of the results obtained from various Mediterranean-type ecosystems. It was not possible to make the coverage exhaustive because some results have been published in journals unavailable to me whilst other results were obtained using methods that made them incomparable with others. Some generalities about community structure and species richness will be attempted, however.

Faunal composition

The first element of community structure that is easily obtained from any study is the presence or absence of certain animal groups. In the case of soil fauna, the minimum taxonomic levels used in the articles reviewed are: phylum for Nematoda, Mollusca and Annelida, class for Myriapoda and order for Crustacea, Arachnida and Insecta. For many groups, the limited number of reports arises not because of their absence from the study area but from the use of selective methods or from the decision of the authors not to include them. For example, many authors do not include microarthropods; some others exclude ants and termites because of their highly contagious distributions. By looking at each site in detail one can note, however, a consistent absence of certain groups (Tables 4.1 and 4.2). For instance, Diplopoda, Thysanoptera and Psocoptera are absent from the two most arid areas studied, viz. the Mariut *Anabasis-Thymelaea* area in Egypt (Cancela da Fonseca *et al.*, 1984; Ghabbour *et al.*, 1977, 1985; Ghabbour & Shakir, 1980) and the asphodel desert in Greece (Pantis *et al.*, 1988).

Relative importance

The relative importance of the groups of soil animals can be estimated using either of two methods. One is to measure population density using

M. Arianoutsou and R.H. Groves, Plant-Animal Interactions in Mediterranean-Type Ecosystems, 37–45, 1994.

Table 4.1. Population densities in individuals/m^2 of soil invertebrates in various Mediterranean-type ecosystems

	A	B	C	D	E	F	G	H	I	J	K	L	M	N
Nematoda											0–93			
Mollusca												1.27		
Annelida												1.16		
Isopoda	0.18	0.03	3.68	5.69	0–3	27–107	17.6	4.4	2.8	6	0–64	2.29	12.5	3.8
Araneae	5.61	3.8	1.24	38.95	29–163	13–91	68.7	38.3	11.3	17.6	28–59	0.63	6.5	4.6
Pseudoscorpiones	0.17	0.08		68.77	43–158		133.5	30.1	14.1	14.3	0–257	1.78	13.9	8
Opiliones				1.36			1.2		0.3	0.3	0–24	0.09		
Scorpiones	0.05	0.03	0.04				0.4					0.02		
Solpugida			0.04											
Acarina	0.11	0.24	0.12			5–224	4966.4	5326.7			37468–121641	66.51	1067.6	803.9
Diplopoda				39.01	3–174	0.112	253.4	44.8	11.1	19	45–149	0.99	5.4	2.5
Chilopoda	0.09	0.02		18.33	0–13		37.1	12.2	0.7	3.7	44–179	0.86	1.3	0.7
Pauropoda											44–1592	0.01	2.2	0.8
Symphyla							2.2	18.8	0.1		306–1617	0.12	2.8	0.3
Collembola							1208.3	4530.4			1615–12788	14.44	190	40.1
Protura							0.3	1.1			87–567	0.04	4.7	3.8
Diplura							17.2	1	0.1		0–131	0.19	0.8	0.3
Thysanura	0.57	1.35	0.24	4.42	0–14	0–43	0.3	0.8	2.9	2.8	0–6	0.07	0.2	0.1
Orthoptera	0.05								1		0–2	0.01		
Dermaptera							0.3	0.2		0.1	0–22			
Embioptera				1.5			1.8	0.5	3.2	3	0–2			
Dictyoptera	5.3	5.27	11.56	16.21	3–16		1.8	1.3	1.9	0.3	0–2	0.06		
Isoptera											0–371			
Psocoptera					3–61	0–91	139	106.6	8.8		0–265	0.19	9.1	17.9
Hemiptera	2.8	3.19	2.56		13–181	37–123	22.8	71.9	5.5	13.7	3–539	0.98	39	16.7
Thysanoptera					61–189	11–21	116	324.5	10		4–105	0.31	4.5	7.8
Neuroptera	5.6	4.59	3.38				0.1	2				0.04		
Lepidoptera			0.6		0.27		1.7	3.2	32.7		20–95	0.53	3.7	3.8
Diptera	0.8	1.03	1.44		11–117	32–603	389.3	54.8	30.2		93–670	1.53	18	5.9
Hymenoptera	15.44	7.29	3.3		4–64	0–5					9–287	0.85	0.8	0.7
Coleoptera	9	5.43	31.5	25.58	13–355	117–464	50.8	112.5	7.4	5.1	408–744	5.03	13.9	8.9

A: *Anabasis* formation in Omayed, Egypt (Cancela da Fonseca *et al.*, 1984), B: *Thymelaea* formation in Omayed, Egypt (Cancela da Fonseca *et al.*, 1984), C: *Thymelaea* formation in Gharbaniat, Egypt (Ghabbour *et al.*, 1985), D: *Pinus halepensis* forest in Sophiko, S. Greece (Karamaouna, 1990), E: *Arbutus unedo* formation in Ikaria is., Greece (Magioris, 1989), F: *Quercus coccifera* formation in Ikaria is., Greece (Magioris, 1989), G: *P. halepensis* forest in Skopelos is., Greece (Radea, 1989), H: *P. halepensis* forest in N. Euboea, Greece (Marmari, 1991), I: *Q. coccifera* formation in Naxos is., Greece (Magioris, 1991), J: *Juniperus phoenicea* formation in Naxos is., Greece (Karamaouna *et al.*, 1993), K: *Eucalyptus marginata* forest in S.W. Australia (Postle, 1985), L: Sclerophyllous forest in Chile. Results expressed as percentages of total density (Di Castri, 1963), M: Sclerophyllous forest in Chile. Results expressed as number/dm^3 (Di Castri & Vitali-Di Castri, 1981), N: Sclerophyllous mattoral in Chile. Results expressed as number/dm^3 (Di Castri & Vitali-Di Castri, 1981).

a variety of techniques (Berlese–Tullgren funnels, Winkler–Moczarski apparatus, hand-sorting, etc.) and the other is to use pitfall traps of various sizes. In order to compare the results from various studies, it was decided for this review to use results obtained as individuals per m^2 for density measurements and individuals per trap per 30 days for pitfall results. All the results used refer to either the average or the minimum and maximum measurements obtained throughout at least one annual cycle.

The most important groups in all areas studied are Coleoptera, Araneae, Hemiptera, Diptera, Diplopoda and Pseudoscorpiones (Table 4.1). To these must be added Acarina, Collembola and ants that, although most probably the most abundant groups, are not always included because they require different methods. The drier areas of Egypt show a high density of Dictyoptera, Neuropteran larvae and Thysanura whilst within Coleoptera, family Tenebrionidae seems to dominate. These areas also show either a very low number or absence of Diplopoda, Chilopoda, Pseudoscorpiones and Psocoptera (Cancela da Fonseca *et al.*, 1984; Ghabbour *et al.*, 1985). These latter four groups have high densities in southwestern Australian forests of *Eucalyptus* (Postle, 1985). The Mediterranean areas of

Table 4.2. Number of animals caught in pitfall traps expressed as individuals/trap/30 days

	A	B	C	D	E	F	G	H	I	J
Mollusca				2						
Annelida					1					
Isopoda	0.4	0.1	0.5	7	11	54	0.1	1.4	6.8	0.2
Amphipoda							0.1	8.9		
Araneae	2	6.8	11.8	353	150	52	1.8	13.2	5.7	6.8
Pseudoscorpiones		0.1	0.4	2	2			0.9	0.1	0.8
Opiliones	2	1.8	6	11	28	22		1.1	0.2	
Scorpiones	0.1			5	2	7			0.1	0.1
Solpugida								1.4	0.5	0.5
Acarina				655	394	1	28.8	11.1		
Diplopoda		0.03	0.8	17	5	1	0.3	1.5	1.8	0.6
Chilopoda	0.9	0.04	0.6	29	7		0.1	0.8	0.1	0.1
Symphyla								0.2		
Collembola							143.2	47.6	0.7	4.9
Diplura	1.5									
Thysanura		0.2				11	1.2		0.2	3.6
Orthoptera		0.5		62	51		0.9	6.5	1.5	2.2
Dermaptera			0.5	2222	580	209	17.4	0.5		
Embioptera		0.04	0.1							
Dictyoptera		2.9	0.4	7	18	46	0.2	1.5		
Psocoptera		0.6					0.1	0.9	0.5	0.3
Hemiptera	3.6	2.2	4.9	65	9	17	0.4	10.8	2.3	3.6
Thysanoptera		0.03		2	1		0.3	0.2	1.9	0.2
Mecoptera				5						
Lepidoptera		0.2		3	8	1		0.3	1.1	1.3
Trichoptera					1					
Diptera		0.4					6.1	60	10.8	18.9
Hymenoptera	4.1			1264	894	693	32	149.3	6.2	7.6
Coleoptera	45.6	2.7		831	785	215	12.1	27.2	15.6	18.9
Siphonaptera			12.8					0.2	0.1	0.1
Strepsiptera								0.3		

A: Asphodel desert in C. Greece (Pantis *et al.*, 1988), B: *Quercus coccifera* formation in Naxos is., Greece (Magioris, 1991), C: *Juniperus phoenicea* formation in Naxos is., Greece (Karamaouna *et al.*, 1993), D: *Eucalyptus marginata* forest in Perth, S.W. Australia (Koch & Majer, 1980; Majer & Koch, 1982), E: *E. marginata* forest in Dwellingup, S.W. Australia (Koch & Majer, 1980; Majer & Koch, 1982), F: *E. marginata* forest in Manjimup, S.W. Australia (Koch & Majer, 1980; Majer & Koch, 1982), G: *Eucalyptus* forest in Mt. Lofty Ranges, S. Australia (Greenslade, 1988), H: Fynbos in Jonkershoek Valley, S. Africa (Donnelly & Giliomee, 1988), I: Xerophytic scrub mattoral in Fray Jorge Natnl. Park, Chile (Saiz, 1988), J: Thorny scrub mattoral in Fray Jorge Natnl. Park, Chile (Saiz, 1988).

Greece are characterized by high densities of Diplopoda, Pseudoscorpiones, Araneae and Coleoptera (Radea, 1989; Karamaouna, 1990; Magioris, 1989, 1991; Marmari, 1991; Karamaouna *et al.*, 1993) whilst in Chile, the dominant groups are Coleoptera, Hemiptera, Diptera, Psocoptera and Pseudoscorpiones (Di Castri, 1963).

Data from pitfall traps show similar tendencies (Table 4.2). Differences occur with the groups that move very slowly such as Diplopoda and hence are under-represented in the traps, or else are very active, such as Dermaptera (in Australia), Orthoptera and Opiliones.

Trophic levels

Looking at the main groups in the various trophic levels, one can see that detritivores are the most diverse. In the arid areas of Egypt, it is mainly the Dictyoptera together with Isopoda, Thysanura and Dipteran larvae that dominate at this level. In southwestern Australia, the dominant detritivore groups are Dipteran larvae, Isoptera and Diplopoda. In the Greek Mediterranean ecosystems and in Chile it is mainly Diplopoda and Dipteran larvae together with Isopoda.

Among the carnivores, Araneae and Pseudoscorpiones are present at high densities in all ecosystems. Neuropteran larvae are very abundant in Egypt whilst Chilopoda can be found in high numbers in certain cases.

Hemiptera seem to be the main phytophagous group in most Mediterranean-type areas. Finally, it must be noted that some groups that exhibit high densities, such as Psocoptera, Hymenoptera and Coleoptera, cannot be assigned to a particular trophic level as they include species with different feeding habits.

The complexity of the soil communities can also be seen in the succession of one group by another at the same trophic level as environmental conditions change. As climate changes from very wet to dry, the dominant macroarthropod decomposers on the island of Naxos, Greece, change from Diplopoda to Isopoda to Thysanura (Karamaouna *et al.*, 1993).

Spatial variation

In the majority of Mediterranean-type ecosystems there is significant spatial variation of the community structure, caused mainly by the heterogeneity of the environment. In the humid and arid areas, at the two extremes of Mediterranean-type ecosystems, this variation is less pronounced. The spatial heterogeneity is both horizontal and vertical. Horizontal heterogeneity is because of the patchiness of an environment that offers areas of open ground, ground under stones, thin or thick litter of various quality, ability to retain water or availability of space, litter under thick or sparse cover, stones in litter under cover, thick or sparse understorey vegetation, etc. Vertical heterogeneity includes the above-ground vegetation, litter, stones and the soil which, at least in calcareous areas, is full of cracks and fissures and may retain in the deeper layers a relative humidity of 100% for most of the year (Ghabbour, 1979). The soil may have a more abundant fauna with higher densities than litter but litter has more species (Postle, 1985; Postle *et al.*, 1991). Bigot and Bodot (1972) have shown that because of the interdependence of invertebrates of the canopy, the litter, the stones and the soil, we cannot really speak of four different communities but of one, characteristic of the dominant plant species.

Spatial variation is accentuated by seasonal variation which causes significant micromigrations from one microhabitat to the other.

Temporal variation

Contrary to what was believed earlier (Di Castri & Vitali-Di Castri, 1981), there does not seem to be a common pattern of temporal variation in community structure in Mediterranean-type ecosystems. If we look at total arthropod densities or biomass, in some areas there are two peaks, one in spring and one in autumn, with a maximum in some areas in the spring and in others in the autumn. In other areas, only one peak is observed and this can be either during winter or summer. There are many reasons for this lack of a common pattern. In Australia, the summer season is by no means as dry as that experienced in similar climatic zones in the Northern Hemisphere. Summer rainfall is erratic and enables the dominant plants to produce new shoots. This may be a relict from a subtropical climate during the Tertiary (Edmonds & Specht, 1981). In areas with mild winters, such as South Africa, many animal groups can produce high densities during winter, correlated with rainfall (Donnelly & Giliomee, 1988).

The temporal pattern becomes more complicated if we look at trophic levels or animal groups. Depending on which trophic level is more abundant, we see different temporal patterns. Herbivores are more closely connected with plant growth rate and therefore are expected to be in higher densities during the peak of the growth season. Decomposers are usually more abundant during the wetter but not the colder months. Therefore, if the summer is not completely dry, they may be active throughout. Predators and parasites depend less on seasons because there is always an amount of food available to them (Edmonds & Specht, 1981).

Individual groups show a large variety of temporal patterns. For example, from pitfall results, spiders in South Australia have one peak in early spring (Greenslade, 1988), in southwest Western Australia one peak in spring (Koch & Majer, 1980), in central and southern Greece one peak in early summer (Karamaouna *et al.*, 1993) or one peak in autumn (Pantis *et al.*, 1988), whilst in S. Africa one peak in spring (Donnelly & Giliomee, 1988). Hemiptera in South Australia have one peak in the summer (Greenslade, 1988), in southwest Western Australia in spring and autumn (Majer & Koch, 1982), in central and southern Greece in spring (Karamaouna *et al.*, 1993) or in summer (Pantis *et al.*, 1988) and in S. Africa in the spring and autumn (Donnelly & Giliomee, 1988).

If we go deeper to the family or species level, temporal differences become more marked. In the central Aegean ecosystem dominated by *Juniperus*, the two peaks of Coleoptera – one in spring and one in the autumn – arise because of two families: Carabidae are abundant during spring and Staphylinidae during autumn. Within the Scarabaeidae, the spring peak was because of the abundant presence of two species whilst the autumn peak was because of the abundance of a third species (Trihas & Legakis, 1991). In the *Eucalyptus* forests of Western Australia, there is a wide array of responses to season, with some species being active throughout the year, some confined to a particular season and some to the cooler, warmer or the moister months (Postle *et al.*, 1991).

It can therefore be concluded that although temporal variation in the total number of soil invertebrates or even of one group may give a general idea of phenology, one has to go to the species or at least to the guild level in order to understand the ecological importance of temporal and spatial variation.

One final note of caution must be mentioned. It is unwise to compare phenologies of quantitative densities with phenologies of trapped animals. As concurrent samples have shown for the island of Naxos, Greece, for some groups there was a difference between peaks of up to two months because of a differential effect that environmental conditions may have on population density and activity (Karamaouna *et al.*, 1993).

Effects of environmental factors

The general trend that is known to occur in soil communities in Mediterranean-climate areas is that organic matter is responsible for spatial variation of the community structure at a given time whilst seasonal variations in temperature and water availability are responsible for temporal population changes (Di Castri & Vitali-Di Castri, 1981). A number of other factors has been shown to play a role in structuring soil communities. Wind action in exposed litter under *Thymelaea* bushes in Egypt causes both dispersal of litter and therefore diminishing organic matter and desiccation of the soil (Ghabbour *et al.*, 1984). The open, uncovered ground in overgrazed maquis ecosystems has both higher temperatures and reduced organic matter (Magioris, 1991). Therefore, in the more arid Mediterranean areas, temperature and water availability play a significant role in the spatial structuring as well as the composition of soil communities.

Interesting conclusions can be drawn by studying the combination of results from pitfall traps and quantitative samples. In winter, high humidity favours high densities, whilst low temperatures decrease mobility. In spring and early summer, low humidity has a negative effect on the presence of many species, whilst increasing temperatures increase their activity (Karamaouna *et al.*, 1992).

Effects of fire

Fire is an important factor in Mediterranean-type ecosystems as it is both a recurring natural phenomenon and a human-induced disturbance. Few studies have been concerned with the effects of fire on the soil invertebrate fauna. The basic conclusions can be summarized as follows.

Total density, species diversity and composition at the class or order level decrease after fire but recover with few exceptions after 2–7 years (Springett, 1971, 1976; Abbott, 1984; Majer, 1980b, 1984, 1985a,b; Sgardelis, 1988).

The soil animals that are more affected by fire are surface-active invertebrates, litter invertebrates (Majer, 1984, 1985a,b), particularly decomposers and especially those associated with the early stages of decomposition (Sgardelis, 1988), fungal feeders and juveniles (Springett, 1979). The impact of burning on these animals has important implications for the rate of nutrient cycling because many of them help regulate decomposition of litter (Majer & Abbott, 1989). The time of recovery of these animals depends on the rate of recovery of the organic horizons and therefore of the vegetation (Sgardelis, 1988).

The fauna may be influenced by direct physical effects of fire but more significantly by longer-term changes in factors such as food availability, shelter and environmental conditions (Majer, 1984).

Fires late in the summer or in the autumn are less detrimental for soil and litter invertebrates than in spring or early summer as most groups are already taking measures to avoid the hot period of the year (Majer, 1980, 1985a,b).

Unburned plants, logs, patches of litter and areas under stones are important refuges for fauna in burnt areas (Whelan *et al.*, 1980; Majer, 1980; Sgardelis, 1988). The phenology of many soil animal groups was still disturbed at least one year after the fire (Sgardelis, 1988).

There are significant differences in reaction to fire between taxa: immediate density reduction, delayed density reduction, temporary absence, density stimulation or no reaction (Majer, 1984). Results obtained from fire impact studies depend significantly on the intensity of fire studied, the type of experimental design or the taxonomic treatment used in the study (Majer, 1985a,b). Therefore, what are needed in these kinds of studies in Mediterranean-type ecosystems are adequate pre-fire data, adequate site replication, samples taken over a long period and animal identifications to species level (Majer, 1985a,b).

Effects of grazing

Although grazing is an important human activity in Mediterranean-type ecosystems, its effect on soil fauna has been little studied. It seems that the basic effect of overgrazing is to reduce densities of most taxa (Magioris, 1991). However, although protection from grazing may induce an increase of density in most taxa, some, especially the detritivores, are negatively affected, thereby allowing the herbivores to be the dominant groups (Cancela da Fonseca *et al.*, 1984; Magioris, 1991). This may be because of the increase of phytomass and the absence of animal dung. Litter fauna is more significantly affected by grazing whilst animals living in soil are less disturbed.

Under conditions of overgrazing, although the Mediterranean-type ecosystems are disturbed, their biota adapt to this stress by using seasonal and annual mechanisms of adjustment (Cancela da Fonseca *et al.*, 1984). Some groups may use both grazed and ungrazed areas during their life cycles. Therefore, a system of periodic or alternate grazing may promote diversity (Magioris, 1991). This partial protection may be more beneficial than complete protection since it favours soil fertility, increases the rate of mineral turnover and does not exaggerate the density of herbivores which may endanger the vegetation in the long run (Cancela da Fonseca *et al.*, 1984).

Effects of agriculture

With the intensification of agricultural practices in Mediterranean-type ecosystems, true detritivores of the original system, capable of dealing with the available plant and animal litter and adapted to the mediterranean climate, are replaced by other mesic detritivores adapted to changes in litter quality (less animal and more plant) and also adapted to the changed physical and chemical soil conditions. Meanwhile, the increase in plant biomass encourages the appearance and proliferation of cryptic phytophages (agricultural pests), occupying the niche of the removed large herbivores. Carnivores also change accordingly (Ghabbour *et al.*, 1985).

Other effects

The effects of mining and subsequent rehabilitation on soil fauna have been studied principally in Mediterranean-type ecosystems in Australia. Highest densities were observed in areas revegetated with a diverse mixture of native plants mainly because of the formation of a species-rich understorey compared with the relatively simple monoculture of revegated areas (Majer & Abbott, 1989).

Trampling by humans in Mediterranean pine forests basically causes both a reduction in population density and changes in the phenology of various groups. The most affected groups are the detritivores, some of which may be completely absent from heavily trampled areas. Most of the predators have lower densities and disturbed phenologies, whilst for taxa such as Dipteran larvae and Collembola, their phenology is not affected. Coleoptera seem to be sensitive indicators of the effects of trampling (Marmari, 1991).

Species richness

The patchiness of Mediterranean-type ecosystems is reflected in the high diversity of soil animal species compared with neighbouring temperate and arid regions. This high diversity has posed problems for the detailed study of these animals because identification to species level is hampered by the lack of specialists in systematics and the cost in time to sort the high number of samples needed to cover environmental heterogeneity. It is therefore not surprising that very few studies have gone to specific level.

Species richness follows in general the spatial and temporal patterns of abundance of soil animals. There are, for example, more species in humid areas than in arid ones (Di Castri & Vitali-Di Castri, 1981; Saiz, 1988). However, although the abundance of invertebrates is higher in soil than in litter, the diversity is higher in litter and diminishes progressively from the surface downwards (Postle, 1985; Saiz, 1988; Postle *et al.*, 1991). Species diversity also shows greater seasonal fluctuations in litter than in soil (Lions, 1977).

Less disturbed sites are richer in species than more disturbed sites and more diverse habitats are richer in species than more uniform ones (Gross, 1985). Sites at high elevations may have more (Di Castri & Vitali-Di Castri, 1981) or less species (Marcuzzi, 1966) than those at lower elevations and coniferous formations such as *Pinus halepensis* may have less species than sclerophyllous formations (Marcuzzi, 1968), although there are cases of coniferous formations having more abundant populations than sclerophyllous formations (Radea, 1989; Marmari, 1991). Lateritic soils may have more species than loamy soils (Postle *et al.*, 1991).

Conclusions

It is still too early to be able to have a detailed view or to construct a functional model of community structure for Mediterranean-type soil fauna. Whole areas of ecological research have yet to be explored, such as trophic interrelationships and connectance, effects of competition and predation on community structure, effects of invasive plants and animals, factors affecting species richness, fluxes of energy, relationships with other components of the mediterranean ecosystems, and many more. The most important element that arises from all recent work is the spatial, temporal and ecological complexity of this system, that can be better understood only by detailed species-level, long-term, seasonal and multi-sample studies and with the cooperation of scientists from all five areas of mediterranean climate in order to exchange information and to establish common sampling, statistical methods and techniques.

References

Abbott, I. 1984. Changes in the abundance and activity of certain soil and litter fauna in the Jarrah forest of Western Australia after a moderate intensity fire. Aust. J. Soil Res. 22: 463–469.

Bigot, L. & Bodot, P. 1972. Contribution à l'étude biocoenotique de la garrigue à *Quercus coccifera* II. Composition biotique du peuplement des invértebrés. Vie Milieu 23: 229–249.

Cancela da Fonseca, J., Ghabbour, S.I. & Hussein, A.K.M. 1984. Characterization of soil mesofauna in a xero-mediterranean ecosystem after a 3-year grazing management. Ecol. Medit. 10: 121–131.

Di Castri, F. 1963. Etat de nos connaissances sur les biocoenoses édaphiques de Chili. In: Doeksen, J. & Van der Drift, J. (eds.), Soil Organisms. North Holland, Amsterdam, pp. 375–385.

Di Castri, F. 1973. Soil animals in latitudinal and topographical gradients of mediterranean ecosystems. In: Di Castri, F. & Mooney, H.A. (eds.), Mediterranean-type Ecosystems. Origin and Structure. Springer Verlag, Berlin, pp. 171–190.

Di Castri, F. & Vitali-Di Castri, V. 1981. Soil fauna of mediterranean-climate regions. In: Di Castri, F., Goodall, D.W. & Specht, R.L. (eds.), Mediterranean-type Shrublands. Elsevier, Amsterdam, pp. 445–478.

Donnelly, D. & Giliomee, J.H. 1988. The epigaeic invertebrate fauna in South African fynbos of different ages after fire. In: Specht, R.L. (ed.), Mediterranean-type Ecosystems. A Data Source Book. Kluwer, Dordrecht, pp. 201–204.

Edmonds, S.J. & Specht, M.M. 1981. Dark Island heathland, South Australia: Faunal rhythms. In: Specht, R.L. (ed.), Heathlands and Related Shrublands of the World. B. Analytical Studies. Elsevier, Amsterdam, pp. 15–20.

Ghabbour, S.I. 1979. Soil invertebrates. In: Ayyad, M.A. & Kassas, M. (eds.), Analysis and Management of Mediterranean Desert Ecosystems. Alexandria Univ., Alexandria, pp. 102–105.

Ghabbour, S.I., Cancela da Fonseca, J., Mikhail, W.Z.A. & Shakir, S.H. 1985. Differentiation of soil mesofauna in desert agriculture of the Mariut region. Biol. Fert. Soils 1: 9–14.

Ghabbour, S.I., Mikhail, W.Z.A. & Rizk, M.A. 1977. Ecology of soil fauna of mediterranean desert ecosystems in Egypt I. Summer populations of soil mesofauna associated with major shrubs in the littoral sand dunes. Rev. Ecol. Biol. Sol 14: 429–459.

Ghabbour, S.I. & Shakir, S.H. 1980. Ecology of soil fauna of mediterranean desert ecosystems in Egypt III. Analysis of *Thymelaea* mesofauna populations in the Mariut frontal plain. Rev. Ecol. Biol. Sol. 17: 327–352.

Greenslade, P. 1988. Density and seasonality of epigaeic fauna in South Australia. In: Specht, R.L. (ed.), Mediterranean-Type Ecosystems. A Data Source Book. Kluwer, Dordrecht, pp. 199–200.

Gross, G. 1985. Seasonality and diversity of epigaeic insects in Adelaide. In: Greenslade, P. & Majer, E.D. (eds.), Soil and Litter Invertebrates of Australian Mediterranean-type Ecosystems. WAIT School of Biology Bull. 12: 43–46.

Karamaouna, M. 1990. On the ecology of the soil macroarthropod community of a Mediterranean pine forest (Sophico, Peloponnese, Greece). Bull. Ecol. 21: 33–42.

Karamaouna, M., Legakis, A., Paraschi, L. & Blandin, P. 1993. Étude d'un écosystème de maquis (île de Naxos, Cyclades, Grèce). Traits généraux du peuplement de macroarthropodes édaphiques. Bull. Ecol. (in press).

Koch, L.E. & Majer, J.D. 1980. A phenological investigation of various invertebrates in forest and woodland areas in the south-west of W. Australia. J. Roy. Soc. West. Austr. 63: 21–28.

Lions, J.C. 1977. Application du concept de la diversité spécifique à la dynamique de trois populations d'Oribates (Acariens) de la forêt de Ste Baume (Var.) 2 pt.: Variabilité temporelle selon les differents nivaux prospectés. Ecol. Medit. 3: 85–104.

Magioris, S.N. 1989. The arthropod fauna of the soil on the island of Ikaria, Aegean Sea, Greece. Biol. Gallo-hellen. 15: 153–158.

Magioris, S.N. 1991. Ecology of soil arthropods in insular phryganic and degraded maquis ecosystems (Naxos is., Cyclades). Doct. Thesis, Univ. of Athens, 209 pp. (in Greek).

Majer, J.D. 1980. Report on a study of invertebrates in relation to the Kojonup nature reserve fire management plan. WAIT School of Biology Bull. 2: 22 pp.

Majer, J.D. 1984. Short-term responses of soil and litter invertebrates to a cool autumn burn in Jarrah (*Eucalyptus marginata*) forest in Western Australia. Pedobiologia 26: 229–247.

Majer, J.D. 1985a. A review of pyric succession of soil and litter invertebrates in south-western Australia. In: Greenslade, P. & Majer, J.D. (eds.), Soil and Litter Invertebrates of Australian Mediterranean-type Ecosystems. WAIT School of Biology Bull. 12: 28–30.

Majer, J.D. 1985b. Fire effects on invertebrate fauna of forest and woodland. In: Ford, J.R. (ed.), Fire Ecology and Management in Western Australian Ecosystems. pp. 103–106.

Majer, J.D. & Abbott, I. 1989. Invertebrates of the jarrah forest. In: Dell, B., Havel, J. & Maluczek, N. (eds.), The Jarrah Forest. A Complex Mediterranean Ecosystem. Kluwer, Dordrecht, pp. 111–122.

Majer, J.D. & Greenslade, P. 1988. Soil and litter invertebrates. In: Specht, R.L. (ed.), Mediterranean-type Ecosystems, a Data Source Book. Kluwer, Dordrecht, pp. 197–226.

Majer, J.D. & Koch, L.E. 1982. Seasonal activity of hexapods in woodland and forest leaf litter in the south-west of Western Australia. J. Roy. Soc. W. Austr. 65: 37–45.

Marcuzzi, G. 1966. Preliminary observations on the soil fauna near Padova (Colli Euganei). Pedobiologia 6: 219–225.

Marcuzzi, G. 1968. Osservazioni ecologiche sulla fauna del suolo di alcune regioni forestali italiane. Ann. Centr. Econ. montana Venezia 8: 209–331.

Marmari, A. 1991. Effect of human activities on the fauna of soil arthropods in a *Pinus halepensis* forest in N. Euboea. Doct. Thesis, Univ. of Athens, 221 pp. (in Greek).

Pantis, J.D., Stamou, G.P. & Sgardelis, S. 1988. Activity patterns of surface ground fauna in Asphodel deserts (Thessalia, Greece). Pedobiologia 32: 81–87.

Postle, A.C. 1985. Density and seasonality of soil and litter invertebrates at Dwellingup. In: Greenslade, P. & Majer, J.D. (eds.), Soil and Litter Invertebrates of Australian Mediterranean-type Ecosystems. WAIT School of Biology Bull. 12: 18–20.

Postle, A.C., Majer, J.D. & Bell, D.T. 1991. A survey of selected soil and litter invertebrate species from the northern Jarrah (*Eucalyptus marginata*) forest of Western Australia, with particular reference to soil type, stratum, seasonality and the conservation of the forest fauna. In: Lunney, D. (ed.), Conservation of Australia's Forest Fauna. Roy. Zool. Soc. NSW, Mosman, pp. 193–203.

Radea, K. 1989. Study of litter, decomposition and the arthropod community in Aleppo pine ecosystems of insular Greece. Doct. Thesis, Univ. of Athens, 256 pp. (in Greek).

Saiz, F. 1988. The composition of soil, litter and epigaeic fauna in three different vegetation associations in the semiarid mediterranean region of Chile. In: Specht, R.L. (ed.), Mediterranean-type Ecosystems. A Data Source Book. Kluwer, Dordrecht, pp. 204–211.

Sgardelis, S. 1988. Effects of fire on the consumers of a phryganic ecosystem. Doct. Thesis, Univ. of Salonica, 221 pp. (in Greek).

Springett, J.A. 1971. The effects of fire on litter decomposition and on the soil fauna in a *Pinus pinaster* plantation. Proc. 4th Coll. Zool. Comm. Int. Soc. Soil Sci. 529–535.

Springett, J.A. 1976. The effect of prescribed burning on the soil fauna and on litter decomposition in Western Australian forests. Aust. J. Ecol. 1: 77–82.

Springett, J.A. 1979. The effects of a single hot summer fire on soil fauna and on litter decomposition in Jarrah (*Eucalyptus marginata*) forest in Western Australia. Austr. J. Ecol. 4: 279–291.

Trihas, A. and Legakis, A. 1991. Phenology and patterns of activity of ground Coleoptera in an insular mediterranean ecosystem (Cyclades is., Greece). Pedobiologia 35: 327–335.

Whelan, R.J., Langedyk, W. & Pashby, A.S. 1980. The effects of wildfire on arthropod populations in Jarrah–*Banksia* woodland. West. Aust. Nat. 14: 214–220.

CHAPTER 5

Bird diversity within and among Australian heathlands

MARTIN L. CODY
Department of Biology, University of California, Los Angeles, CA 90024, USA

Key words: birds, heathlands, diversity, turnover, shrublands, mallee, acacia scrub, mulga

Abstract. Regional bird diversity is higher in northeastern than in southwestern Australia, owing to the larger area of forests in the northeast and its proximity to New Guinea, a source of speciation via repeated invasions. In protead heathlands bird species are autochthonously produced, and the habitat is more common in southwestern and southern Australia than in the northeast. Bird α-diversity (within-habitats) is the same in southwestern and northeastern heathlands, but β-diversity (among habitats) is higher in the southwest, and γ-diversity (between sites) is higher over shorter distances in the southwest. There are more peripheral species in the northeastern heathlands, and these are drawn primarily from adjacent woodlands, although rainforest and sclerophyll forest species also invade opportunistically.

In the southwest, α-diversity increases from protead heathlands through mallee and acacia-scrub to mulga, corresponding to a gradient of increasing productivity largely under edaphic control. The densities of nectarivores decline and those of insectivores increase along this gradient. The core species in the bird community become more predictable from protead heaths to mulga. Soil nutrient status among sites is correlated to leaf specific weight in *Acacia*, which is negatively correlated to overall bird density.

Introduction

This chapter describes patterns of bird diversity in one of Australia's distinct habitat types, protead-dominated heathlands (George *et al.*, 1979; Specht, 1979), and contrasts the patterns to those found in related shrublands dominated by *Eucalyptus* spp. (mallee), *Acacia* spp. (mixed acacia-scrub) and mulga (*Acacia aneura*). The chapter presents a quantitative extension of an earlier review (Kikkawa *et al.*, 1979). The patterns are presented and analyzed in terms of diversity components: within-habitat or α-diversity, between-habitat turnover or β-diversity, and within-habitat, between-site turnover, or γ-diversity. In an earlier paper (Cody, 1993) I conducted a similar analysis of bird diversity in Australian rainforests in the northeast, *Eucalytus*-dominated woodlands and forest (northeast and southwest), and mulga (across central Australia), to examine the effects of habitat area, continuity or fragmentation on species numbers within habitats and species turnover among habitats and sites. This chapter adopts the same approach for heathlands, which are rather limited in overall area, generally coastal in distribution, and occur as local, isolated patches of varying extent on sites with the poorest soils.

In general, regional diversity in birds is higher in northeastern Australia, and substantially lower in the southwest, particularly in forest birds. This appears to be because of two sorts of effects: a) historical/geographical effects such as higher speciation in the northeast via repeated invasions from New Guinea (Keast, 1961); and b) greater isolation and smaller woodland/forest habitat areas in the southwest (Keast, 1985; Cody, 1993). Differences in regional diversity result in higher overall α-diversity in northeast compared to the southwest, by some 8 species per site; β- and γ-diversities are also higher in the northeast. Because protead heathlands are populated largely by the products of autochthonous speciation within Australia, and because the heathlands are more common in the southwest, it might be expected that the diversity patterns in the heathland avifauna would counter these general regional trends.

M. Arianoutsou and R.H. Groves, Plant-Animal Interactions in Mediterranean-Type Ecosystems, 47–61, 1994.

Fig. 5.1. Map of heathland vegetation in Australia (from Specht, 1981), showing locations of study sites in SW and NE Australia. EQ: Gunbarrel Hwy.; EF: Boorabbin N.P.; FN: Condingup; FM: Cape Le Grande N.P.; FL: Munglinup; Fl: Stirling Ranges N.P.; FD: William's Bay N.P.; DM: Yanchep N.P.; DL: Cervantes; DK: Badgingarra N.P.; CL: Heathlands; CJ: Tozer's Gap; BN: Deepwater N.P.; Cl: Marcus Beach.; CP: Emu Creek; CB: Dave's Creek; CD: Broadwater N.P.

Protead heathlands

Study sites and field methods

Protead heathland is low (to about 6 m) and rather dense vegetation dominated by shrubs in the family Proteaceae (many species and genera, but especially *Banksia*), and occurs discontinuously around coastal Australia from the southwest along the southern coast and up the eastern coast. There is a major break in the Nullabor region (southwest); the heathlands are generally more extensive in the southwest, and become extremely localized in the northeast. Heathlands are found on a range of nutrient poor (especially phosphorus poor, <0.002%; Groves, 1981) basement, laterite and sand-over-laterite soils, and are replaced by *Eucalyptus*-dominated mallee on richer soils, and by a mixed *Acacia* scrub on Quaternary sands. In turn, on red earth soils in the drier (<250 mm annual precipitation) interior, a bushland of *Acacia aneura* (mulga) replaces the heathlands. Net productivity in mulga is nearly twice that of mallee, and twice again that of protead heaths (Specht, 1979; Sattler, 1986).

Heathland habitats were censused for bird species and densities in southwestern Australia (1984–85) and northeastern Australia (1989–90),

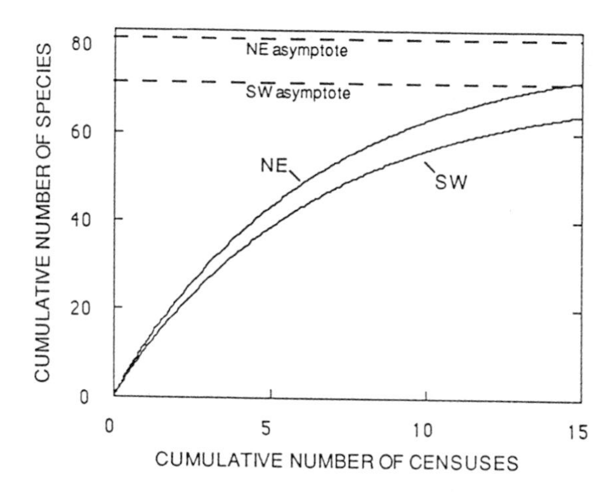

Fig. 5.2. Cumulative numbers of species from censuses assembled in random order, NE and SW Australia. The fitted exponential curves have significantly different asymptotes, 80 species in the NE, 71 in the SW.

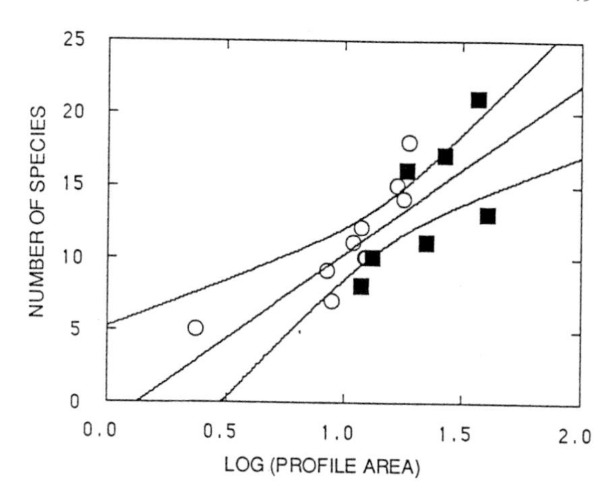

Fig. 5.3. Number of species (α-diversity) of protead heath sites as a function of vegetation structure (LOGAREA = log[area under foliage profile]). Shaded squares: NE sites; open circles: SW sites.

using standard census techniques in c. 5 ha sites within contiguous and undisturbed vegetation (Cody, 1993). The springtime censuses were of diurnal, non-raptorial, non-parasitic and non-aerial feeding birds only. In all, 154 sites were censused, in a wide variety of habitat types. Of heathland sites dominated by proteads (PH), I employed 10 in the southwest and 7 in northeastern Australia (see Fig. 5.1). In southwestern Australia, censuses were conducted also in other heathland habitats, including 8 habitats dominated by mallee eucalypts and 7 dominated by acacias; 6 census sites of non-heathland mulga (*Acacia aneura*) scrubland are also included in this comparison.

At each site I recorded plant species composition and cover, and the foliage profile of the vegetation by plotting vegetation density against height above ground. The foliage profiles average 20 sightings through the vegetation, ground to canopy, along four orthogonal directions at each of five haphazardly selected points within each site. At many census sites *Acacia* species were collected, identified and preserved, and used to generate values of *Acacia* 'leaf' specific weights (phyllode dry weight per unit area, g/cm^2).

Species turnover between any pair of sites is measured by the formula

$$[1-C(T_1+T_2)/2T_1T_2]=\text{SPTURN},$$

where site species totals are 'T_1' and 'T_2' and the species in common between sites is 'C'. Species turnover might be generated by differences in resources or foraging opportunities, and estimated by differences in vegetation structure (from foliage profiles) as VGTURN; this component of species turnover is β-diversity. It might also be related to distance apart of the census sites, DIST, in which case turnover measures γ-diversity. These two diversity components are potentially independent. The first is controlled by vegetational differences between sites; the other component is controlled by the sites being located in different regions with different avifaunas, the latter being due to species replacements on a geographical scale and to their being adjacent to different vegetation types in the different regions (and hence subject to invasion by a different set of peripheral species).

Alpha-diversities in protead heathland sites

In the 17 protead heathland (PH) sites, some 54 bird species were found in the NE, and 57 in the SW. The mean α-diversity of PH sites is 12.4 ± 3.9 species in the NE, and 11.1 ± 3.8 in SW sites; thus mean α-diversity is virtually identical between regions (nonsignificant statistical difference). By randomizing the censuses and plotting cumulative species graphs, it is predicted that

Table 5.1. Characteristic species of Australian protead heathlands

1. Species found only in protead heathlands	
White-streaked honeyeater (NE) (*Trichodere cockerelli*)	Rufous-banded honeyeater (NE) (*Conopophila albogularis*)
New Holland honeyeater (NE) (*Phylidonyris novaehollandiae*)	Red-backed buttonquail (NE) (*Turnix maculosa*)
2. Species reaching highest density in protead heathlands	
Little wattlebird (NE) (*Anthochaera chrysoptera*)	Variegated wren (NE) (*Malurus lamberti*)
Brown honeyeater (NE, SW) (*Lichmera indistincta*)	Southern emuwren (NE) (*Stipiturus malachurus*)
Singing honeyeater (SW) (*Lichenostomus virescens*)	Tawny grassbird (NE) (*Megalurus timoriensis*)
White-cheeked honeyeater (NE, SW) (*Phylidonyris nigra*)	Grey shrikethrush (NE) (*Colluricincla harmonica*)
Tawny-crowned honeyeater (SW) (*Phylidonyris melanops*)	Western thornbill (SW) (*Acanthiza inornata*)
Brown-backed honeyeater (NE) (*Ramsayornis modestus*)	Scarlet robin (SW) (*Petroica multicolor*)
White-fronted honeyeater (SW) (*Phylidonyris albifrons*)	Field wren (SW) (*Calamanthus fuliginosus*)
White-tailed black cockatoo (SW) (*Calyptorhynchus baudinii*)	Little corella (SW) (*Cacatua sanguinia*)
3. Species reaching $\geq$50% maximum density in $\geq$50% of sites	
Torresian crow (NE) (*Corvus orru*)	Australian raven (SW) (*Corvus coronoides*)
4. Species present in $\geq$50% of sites	
Grey-breasted white-eye (NE, SW) (*Zosterops lateralis*)	Red-backed wren (NE) (*Malurus melanocephalus*)

PH censuses in the northeast would collect a total of 80 species (S_T, the asymptote of the cumulative species graph, Fig. 5.2), and at southwestern sites 71; in each region, 15 censuses would be required to reach 90% of the predicted asymptote S_T. The difference between regions in predicted S_T, viz. 9 species, is very close to the species number by which northeastern sites exceed southwestern sites in α-diversity overall (8.5 species; Cody, 1993).

I next examine whether variations in α-diversity are attributable to a difference in vegetation structure among sites (see, e.g., Recher, 1985; Gilmore, 1985, for Australian forest and woodland birds). For both northeastern and southwestern sites log (profile area) is a significant predictor of species α-diversity (whereas log[vegetation height] is not). A linear regression through the origin (Fig. 5.3) accounts for most of the variance in species numbers in both regions. The slope of this line does not differ significantly from the northeast to the southwest, nor from the regression line that passes through the origin:

$$\text{SPECIES} = 10.532^{*}\text{LOGAREA}, \quad p<0.001, \quad r = 0.980.$$

The bird species of protead heathlands

There are few bird species restricted to protead heathlands, but some species are characteristic of this habitat (listed in Table 5.1). Honeyeaters (Meliphagidae) in general and species of *Phylidonyris* in particular are common, with 10 species in the family and four in the genus being regular in PH sites. In all, 25 meliphagids were recorded, but since many species have restricted geographic ranges just four occur in both northeastern and southwestern censuses. Ground-foraging insectivorous malurid wrens are reasonably common (7 species, NE + SW), but foliage insectivores are

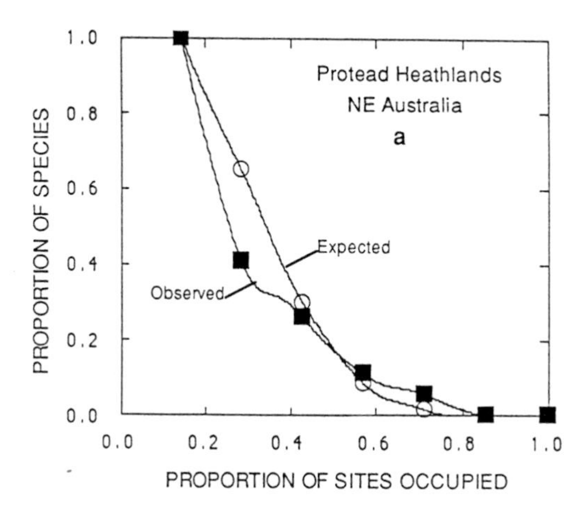

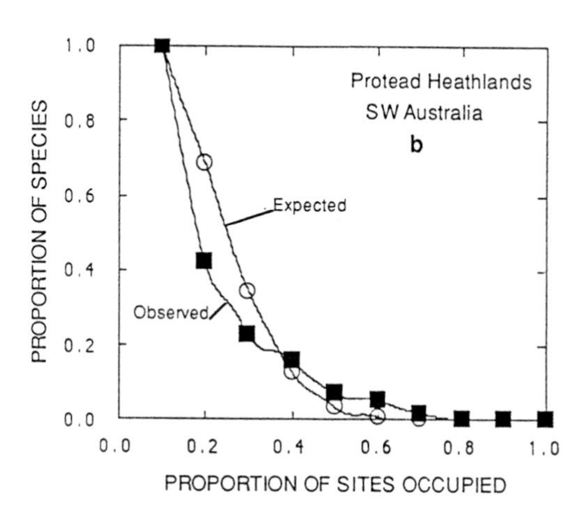

Fig. 5.4. The cumulative proportion of total heath species (ordinate) is plotted as a function of the proportion of sites in which they occur (abscissa). Shaded squares: observed data; open circles: predicted from binomial model of random allocation of species among sites. Both regions show rather more common species, and considerably more rare species, than the model predicts.

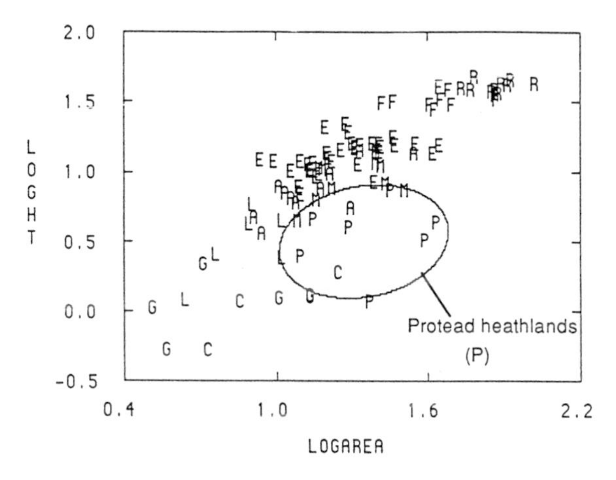

Fig. 5.5. Census sites in NE Australia are plotted in the habitat plane of LOGHT (log[vegetation height]) and LOGAREA (log[area under foliage profile]). The letter symbols represent R: rainforest; F: sclerophyll forest; E: woodland; L: mallee-spinifex; G: grassland; C: chenopod shrubland; A: acacia scrub; M: mulga and P: protead heathland. The last habitat is also depicted by its 90% contour ellipse.

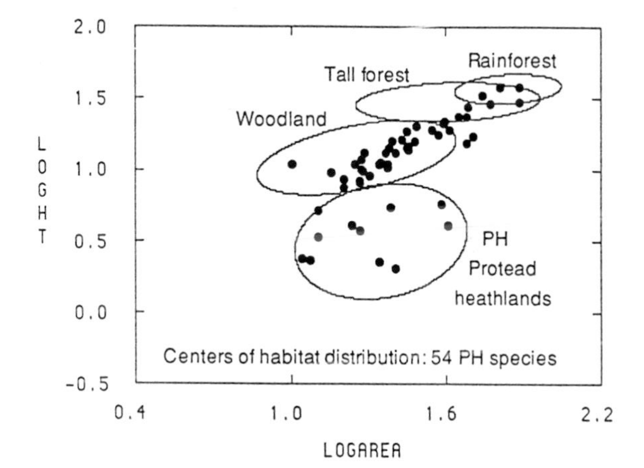

Fig. 5.6. Points in the habitat plane represent the center of habitat distribution, weighted by density, of each of the 54 species that occur in PH habitat in NE Australia. The contour ellipses of rainforest, tall forest and woodlands, as well as of PH, are shown for reference.

rare, with just one species (*Zosterops pallidus*) being regular (though in low density).

Are there core species in protead heathlands?

For repeated censuses within a habitat type, some proportion of the total species S_T occurs in all sites, a higher proportion in (at least) all sites except one, except two, etc. until all species are accounted for. Such plots can be used to assess the predictability of species identities in the censuses, and observations compared to what is expected, with binomial models, if sites 'sample' or 'collect' species by chance (see Cody, 1986). The binomial probability p is calculated as s_α/S_T, where s_α = mean α-diversity, and the model computes the expected number of "successes" or sites

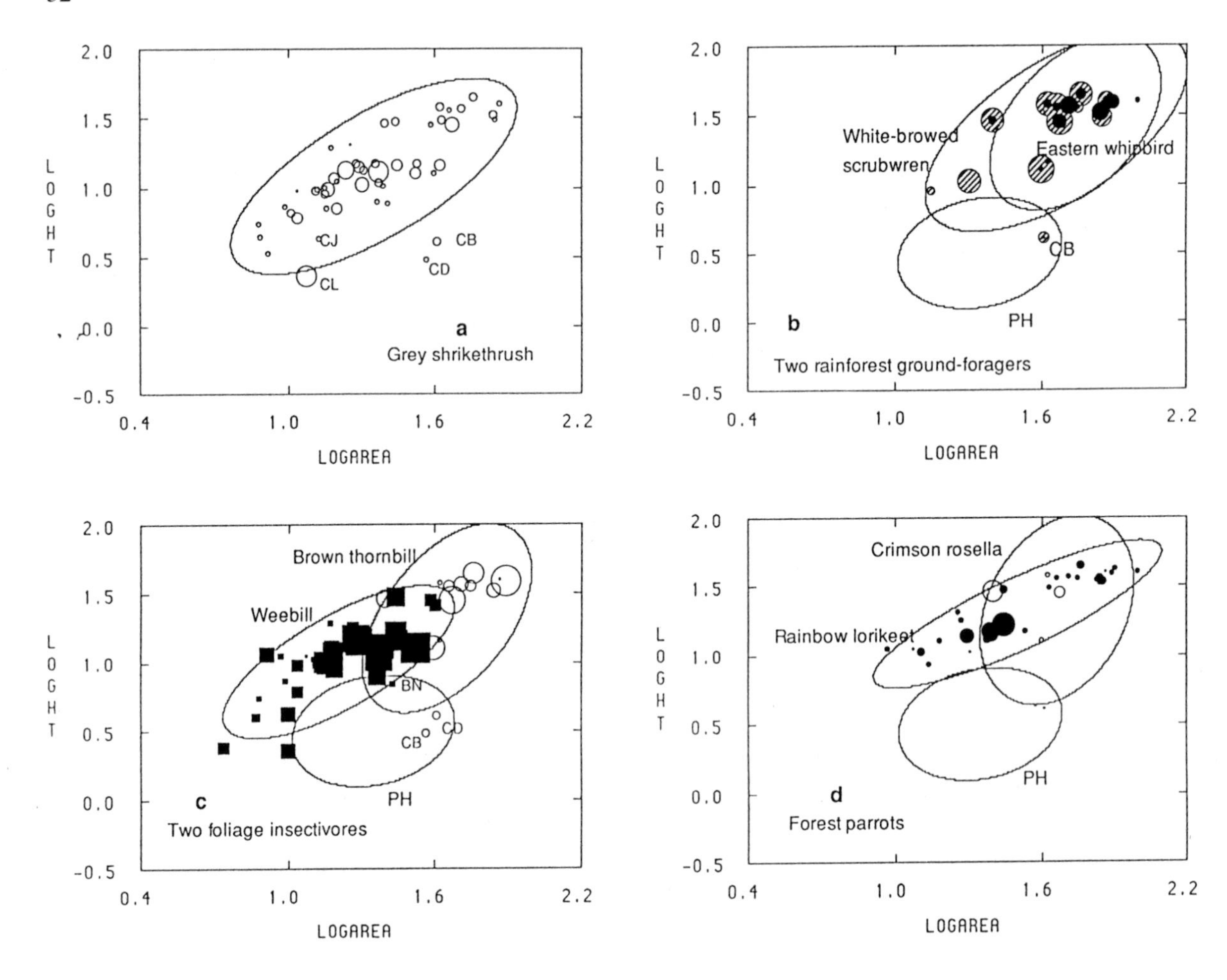

Fig. 5.7. (a) Habitat distribution of Grey shrikethrush in NE Australia. The species occurs in 4 PH sites as indicated. In both Fig. 5.7 and 5.8 species' distributions are characterized by 90% confidence ellipses, and symbol size is proportional to the species' density at each site; (b) Habitat distributions of two ground-foraging insectivores, Eastern whipbird (shaded circles) characteristic of rainforest and White-browed scrubwren (hatched circles) of rainforest and tall sclerophyll forest. Each species occurs at just one PH site, Dave's Creek, a montane heath that is surrounded by rainforest; (c) Habitat distributions of two species of foliage insectivores, Weebill (shaded squares) common in woodlands and Brown thornbill (open circles) typical of tall forest and rainforest. The former occurs in two PH sites, the thornbill in just one; (d) Habitat distributions of two parrots of tall forest and rainforest, Rainbow lorikeet (shaded circles) and Crimson rosella (open circles). Each species occurs in a single PH site.

k in which species are recorded in the '*n*' repeated censuses.

I use binomial models for both northeastern and southwestern PH sites to compare 'observed' with 'expected' distributions (Fig. 5.4); in both the northeast and southwest, there are both too many common (predictable) species and, more disparately, too many rare species to fit the random model. This result holds ($p<0.001$ by Chi-square), whether a zero category is omitted (using actual S_T, whence $s_\alpha/S_T = p = 0.254$, NE, and $s_\alpha/S_T = p = 0.195$, SW) or whether it is included (using the predicted species total S_T, whence $s_\alpha/S_T = p = 0.172$, NE, and $= 0.157$, SW). Despite a weak core of species that are found more frequently than chance dictates, the identities of PH species are in general poorly predictable, and opportunistic honeyeaters as well as stragglers from other habitats form a large component of PH communities.

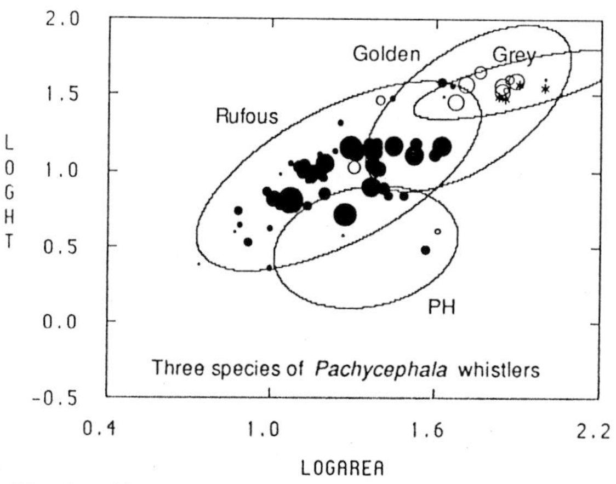

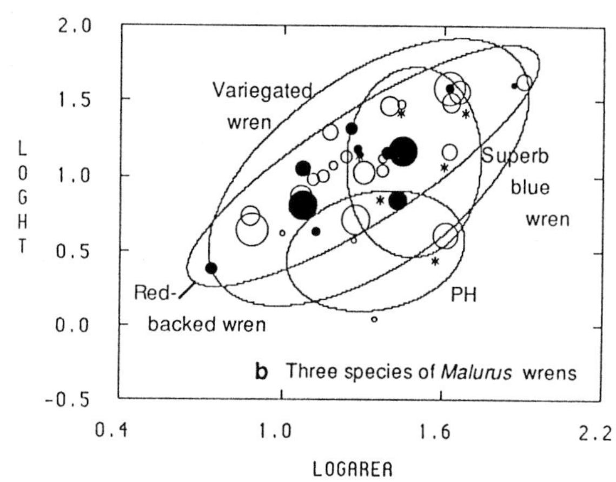

Fig. 5.8. (a) Habitat distributions of three species of *Pachycephala* whistlers showing extensive habitat segregation: Grey (asterisks), Golden (open circles) and Rufous whistler (shaded circles). There are single PH records for the first two species, but several for the Rufous whistler, which is fairly common in heath habitats; (b) Habitat distributions of three species of malurid wrens, showing wide interspecific overlaps: Superb blue wren (asterisks), Variegated wren (open circles), and Red-backed wren (shaded circles). All three species occur in PH sites, the last two regularly, and the variegated wren in PH at its maximum density.

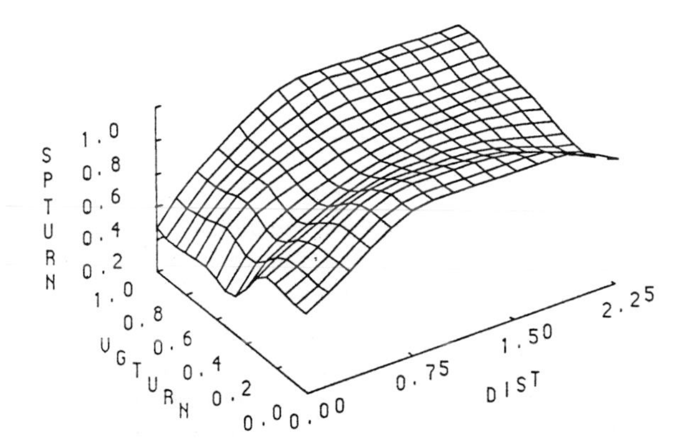

Fig. 5.9. Bird species turnover among sites in NE Australia (SPTURN, ordinate) is represented as a function of two independent variables, distance between sites (DIST, in units of 1000 km) and difference in vegetation structure (VGTURN, distance between points [sites] of the vegetation plane of LOGAREA and LOGHT, as in Fig. 5.5). The surface is fitted to data using the Distance Weighted Least Squares Smoothing method.

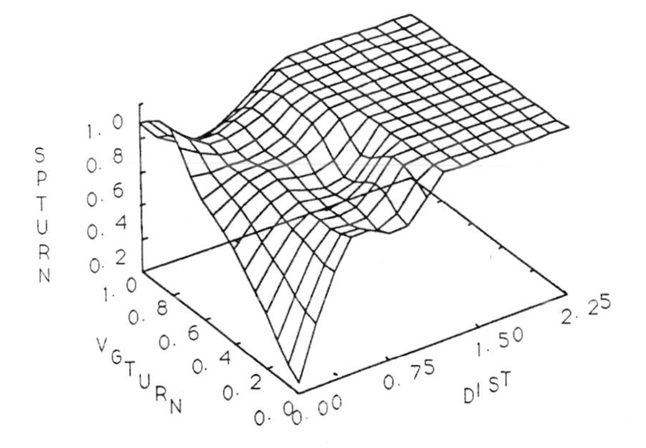

Fig. 5.10. Species turnover in SW Australia (cf. Fig. 5.9). Note low turnover when VGTURN and DIST are both small, and steep turnover with increasing VGTURN and DIST away from the origin.

Where are the peripheral species coming from?

Protead heaths are edaphically restricted to isolated patches among more continuous matrices of other habitat types, frequently woodland and forest. An analysis of the less frequently encountered (peripheral rather than core) heathland bird species shows that they are shared with a wide range of other habitats in which they are more regular in occurrence and in which they reach higher densities. In the NE, these habitats include tropical and subtropical rainforests, Eucalyptus-dominated forests and woodlands.

I examine the habitat affiliations of all bird species recorded in north-eastern protead heathlands by plotting their distributions in the habitat plane defined by log(vegetation height) and log(profile area), LOGHT and LOGAREA

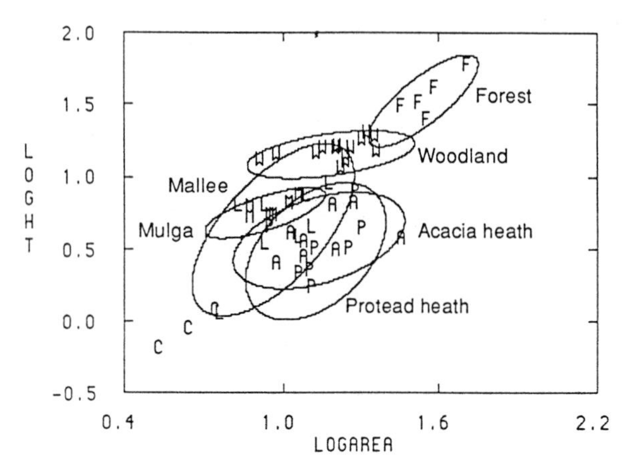

Fig. 5.11. Distribution in the habitat plane of SW Australian habitats. At the top right are forest (F) and, below, woodland (W) habitats. The smaller (center, left) ellipse encompasses mulga (M) habitat, overlapping with, from left to right, mallee-spinifex (L), acacia-scrub (A), and protead-heathland (P). Chenopod shrublands (C) occur in the lower left of the plane (no ellipse).

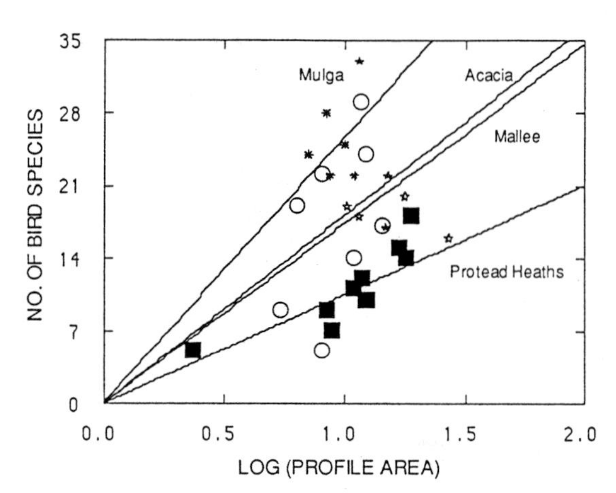

Fig. 5.13. Species number (α-diversity) increases significantly with LOGAREA, a measure of vegetation structure, in PH habitats (filled squares), but the regression lines for mulga (asterisks), acacia-heath (open circles) and mallee (open stars) are not significant.

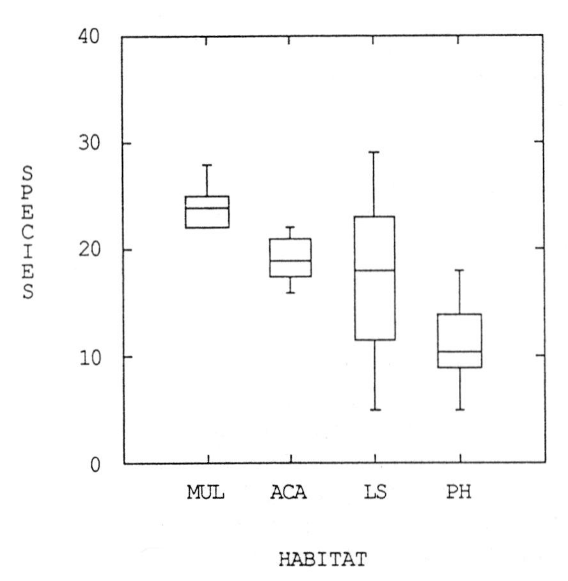

Fig. 5.12. Species number (α-diversity) in four SW habitats, lowest in protead-heathland (PH), and highest in mulga (MUL).

respectively. The species' distributions are defined by contour ellipses in the habitat plane, centered in what can be regarded as prime habitat for the species; the distributions are plotted using census data (occurrence) weighted by site-specific bird density. Fig. 5.5 shows the location of all northeastern census sites in the habitat plane, and the contour ellipse of the PH sites is also indicated. In Fig. 5.6, the centers of distribution of all 54 species that occur in northeastern PH habitats are shown, relative to the contour ellipses for rainforest sites, eucalypt forest sites, eucalypt woodland sites, and protead heathland sites. The analysis shows that ten bird species have density-weighted habitat ranges that are based (centered) within the PH habitat range:

Southern emuwren *Stipiturus malachurus* White-cheeked honeyeater *Phylidonyris nigra* Golden-headed cisticola *Cisticola exilis* White-streaker honeyeater *Trichodere cockerelli* Rufous-banded honeyeater *Conopophila albogularis* Red-backed button-quail *Turnix maculosa* Brown-backed honeyeater *Ramsayornis modestus* Little wattlebird *Anthochaera chrysoptera* New Holland honeyeater *Phylidonyris novaehollandiae* Tawny grassbird *Megalurus timoriensis*

Other and more peripheral species are based in rainforests (three species: Grey whistler *Pachycephala griseiceps*, Eastern whipbird *Psophodes olivaceus*, Green-winged pigeon *Chalcophaps indica*), and six species in tall sclerophyll forest (White-browed scrubwren *Sericornis frontalis*, Yellow-spotted honeyeater *Meliphaga notata*,

Table 5.2. Habitat base of heathland and bushland bird species in southwestern Australia

Mulga-based species:	
Splendid wren (*Malurus splendens*)	Spiny-cheeked honeyeater (*Acanthagenys rufogularis*)
Galah (*Cacatua roseicapilla*)	Rufous whistler (*Pachycephala rufiventris*)
Red-capped robin (*Petroica goodenovii*)	Redthroat (*Sericornis brunneus*)
Crested pigeon (*Ocyphaps lophotes*)	Chestnut-rumped thornbill (*Acanthiza uropygialis*)
Wedgebill (*Psophodes cristatus*)	Grey shrikethrush (*Colluricincla harmonica*)
White-browed babbler (*Pomatostomus superciliosus*)	Jacky winter (*Microeca leucophaea*)
Crested bellbird (*Oreoica gutteralis*)	Slate-backed thornbill (*Acanthiza robustirostris*)
Mulga parrot (*Psephotus varius*)	Southern whiteface (*Aphelocephala leucopsis*)
Cinnamon quailthrush (*Cinclosoma cinnamomeum*)	Zebrafinch (*Poephila guttata*)
Little crow (*Corvus bennetti*)	Yellow-rumped thornbill (*Acanthiza chrysorrhoa*)
Acacia-based species:	
Variegated wren (*Malurus lamberti*)	Grey-breasted white-eye (*Zosterops lateralis*)
Grey fantail (*Rhipidura fuliginosa*)	Singing honeyeater (*Lichenostomus virescens*)
Broad-tailed thornbill (*Acanthiza apicalis*)	Pied honeyeater (*Certhionyx variegatus*)
Australian magpie (*Gymnorhina tibicen*)	White-browed scrubwren (*Sericornis frontalis*)
Blue and white wren (*Malurus leucopterus*)	White-tailed warbler (*Gerygone fusca*)
Mallee/spinifex-based species:	
Budgerigah (*Melopsittacus undulatus*)	Black-faced woodswallow (*Artamus cinereus*)
Weebill (*Smicrornis brevirostris*)	Port Lincoln parrot (*Barnardius zonarius*)
Masked woodswallow (*Artamus personatus*)	Crimson chat (*Ephthianura tricolor*)
Striated pardalote (*Pardalotus striatus*)	Brown-headed honeyeater (*Melithreptus brevirostris*)
Cockateel (*Nymphicus hollandicus*)	Grey currawong (*Strepera versicolor*)
Yellow-rumped pardalote (*Pardalotus xanthopygus*)	Black honeyeater (*Certhionyx niger*)
New Holland honeyeater (*Phylidonyris novaehollandiae*)	
Protead heath-based species:	
Red wattlebird (*Anthochaera carunculata*)	White-eared honeyeater (*Lichenostomus leucotis*)
Brown honeyeater (*Lichmera indistincta*)	White-rumped miner (*Manorina flavigula*)
White-fronted honeyeater (*Phylidonyris albifrons*)	Western spinebill (*Acanthorhynchus superciliosus*)
White-cheeked honeyeater (*Phylidonyris nigra*)	Tawny-crowned honeyeater (*Phylidonyris melanops*)
Western thornbill (*Acanthiza inornata*)	Southern emu-wren (*Stipiturus malachurus*)

Table 5.2. Continued.

Cosmopolitan species:	
Grey butcherbird (*Cracticus torquatus*)	Black-faced cuckooshrike (*Coracina novaehollandiae*)
Willie wagtail (*Rhipidura leucophrys*)	Rufous songlark (*Cinclorhamphus mathewsi*)
White-winged triller (*Lalage sueurii*)	Emu (*Dromaius novaehollandiae*)
Australian raven (*Corvus coronoides*)	Brush bronzewing (*Phaps elegans*)
Magpielark (*Grallina cyanoleuca*)	Richard's pipit (*Anthus novaeseelandiae*)
Southern scrubrobin (*Drymodes brunneopygia*)	Common bronzewing (*Phaps chalcoptera*)

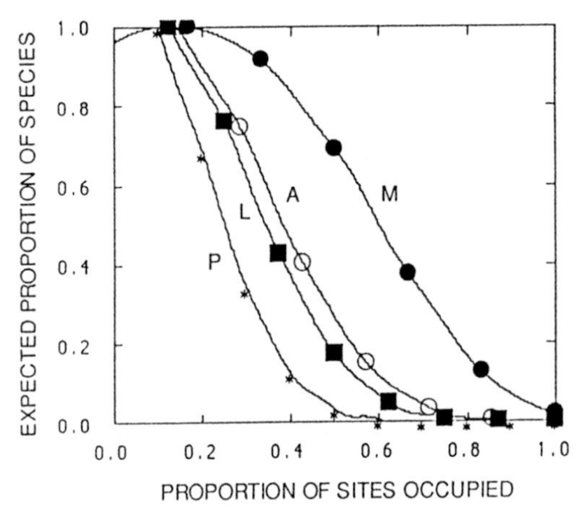

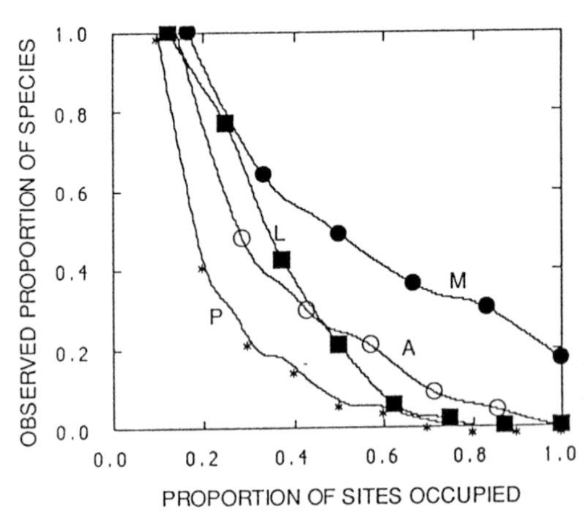

Fig. 5.14. Species predictability in different habitat types in SW Australia (see Fig. 5.4) is tested with binomial models. Expectations from the models are shown at left, and observations (from the census data) at right. There are too many common core species in mulga, and too many peripheral and rare species in acacia- and protead-heath to fit the model. See text for further discussion.

Graceful honeyeater *Meliphaga gracilis*, Golden whistler *Pachycephala pectoralis*, Cicadabird *Coracina tenuirostris*, Eastern yellow robin *Eopsaltria australis*). Two species are predominantly grassland inhabitants, and the remaining 33 species – the numerically dominant component of the PH species list – have habitat ranges centered in eucalypt woodlands.

The distributions of some species that occur marginally in northeastern protead heathlands are given in Fig. 5.7. Grey shrike thrush (*Colluricincla harmonica*, Fig. 5.7a) is a species of particularly wide habitat (and geographical) distribution, and reaches its highest densities in woodlands. The species was recorded at four PH sites, in just one of which (the isolated and species-poor site 'Heathlands' on the Cape York peninsula; see Fig. 5.1), was it common. The distributions of two ground-foraging birds typical of rainforest and tall forest are shown in Fig. 5.7b; both occur in a single PH site, Dave's Creek, a montane heathland surrounded by rainforest in Lamington National Park. Fig. 5.7c shows a woodland foliage insectivore, Weebill *Smicrornis brevirostris*, and a forest species, Brown thornbill *Acanthiza pusilla*, both being extremely marginal in heathlands, with one and two site records respectively. Two forest parrots, Crimson rosella *Platycercus elegans* and Rainbow lorikeet *Trichoglossus haematodus*, are similarly marginal in the heathlands (Fig. 5.7d), with single records for each species at Dave's Creek and Marcus Beach (southern coastal Queensland) respectively.

Three species of whistler (*Pachycephala*) are

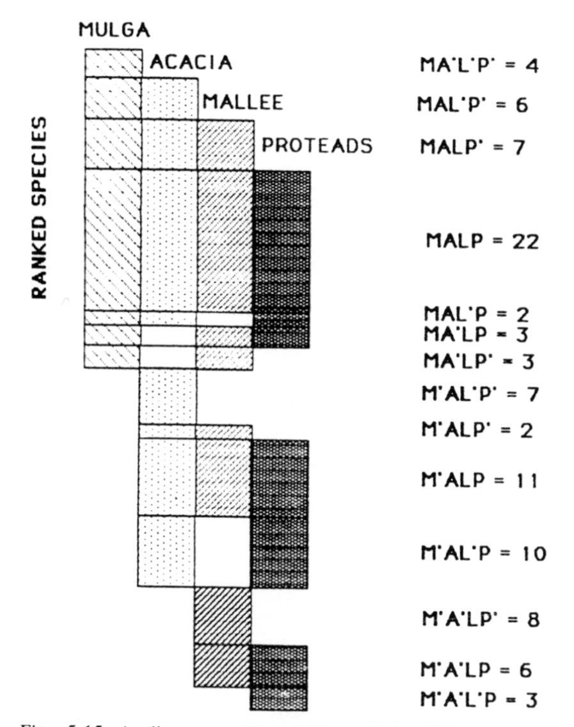

Fig. 5.15. A diagrammatic ranking of the 94 bird species found in four SW habitats, showing species habitat specificity and turnover among habitats. Species number in various habitat distribution categories are shown at the right, where X indicates that a species is found in habitat X, and X′ indicates that it is not (M = mulga, A = acacia-heath, L = mallee- spinifex, and P = protead heath). Joint habitat occupancy of all four habitat types is thus represented 'MALP', a category with 22 bird species.

found in the NE censuses, with habitat distributions generally segregated from rainforest to tall forest to woodland as shown in Fig. 5.8a. All three species occur in the PH sites, but with single records for the rainforest-centered Grey whistler (at Tozer's Gap, adjacent to the Iron Ranges rainforest) and the forest-centered Golden whistler (at Dave's Creek). The woodland-based Rufous whistler *Pachycephala rufiventris*, however, occurs in 3/7 PH sites, in two of which it is common (0.59, 0.57, 0.13 indiv./ha respectively). Finally in Fig. 5.8b I show the distributions of malurid wrens, of which three species show extensive habitat overlaps and are routine inhabitants of PH vegetation. All but one PH site yielded wren species, and 3 sites had two species; Red-backed wren (*Malurus melanocephalus*) and Variegated wren (*M. lamberti*) each occurred in four sites, with Superb Blue wren (*M. cyaneus*) found at a single site. Variegated wrens reach maximal density in PH, and red-backed wrens one-half maximal density in this habitat.

Species turnover among sites: vegetation structure vs. distance

In the PH habitats of northeastern Australia, I find no regular species turnover with vegetational differences among sites; although adjacent PH sites may show a substantial species turnover (c. 50%), this is not enhanced by increasing vegetational differences between the sites. In Fig. 5.9 species turnover (ordinate) is plotted as a function of two independent variables, inter-site distance and difference in vegetation structure. Species turnover increases with distance apart of sites to near 100% turnover at about 800 km, and distant sites show only modestly increasing turnover as the difference in vegetation structure between them increases. In the southwest, in contrast, PH habitats that are both close and structurally similar in vegetation support similar species (near-zero turnover; Fig. 5.10). Keeping inter-site distance small, as VGTURN increases so does SPTURN, to nearly 100% between the most distinct of the PH sites. SPTURN also increases with DIST, steeply for the first 500 km of inter-site separation. Between sites that are 500 km or more apart, VGTURN has relatively little influence in further increasing SPTURN.

I interpret the differences in species turnover patterns between regions as follows: a) In the southwest, where PH habitat is overall more common, and is less fragmented in larger blocks, species predictability is higher, with the same species found in nearby sites of similar vegetation structure; b) SPTURN increases regularly with increasing DIST and increasing VGTURN in the southwest; β-diversity is high in a local area, and γ-diversity is high since inter-site distance correlates well with species turnover between sites up to 500 km apart; c) in the northeast, with PH habitats much rarer and more localized in smaller and isolated patches that are either montane or strictly coastal, there is much less local species predictability in the PH bird community, with a high local turnover among sites because of a larger component of opportunistic species; d) there is no significant β-diversity component locally in the NE, where the same species suite can

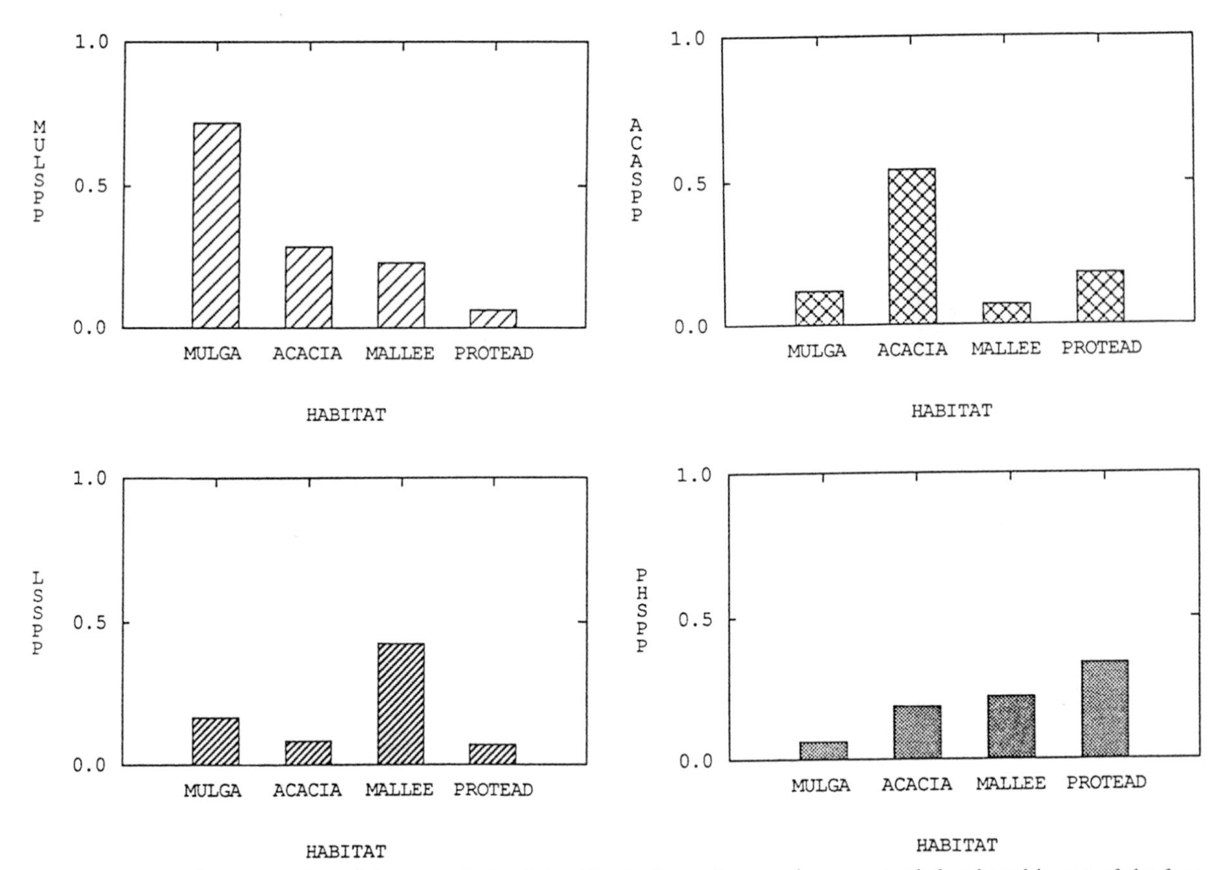

Fig. 5.16. Depending on number of sites occupied and densities in these sites, species are regarded as based in one of the four habitat types. The representation of the mulga-based species attenuates from left to right, and that of protead-heath species from right to left, indicating that species perceive this arrangement of the habitats as a gradient in suitability.

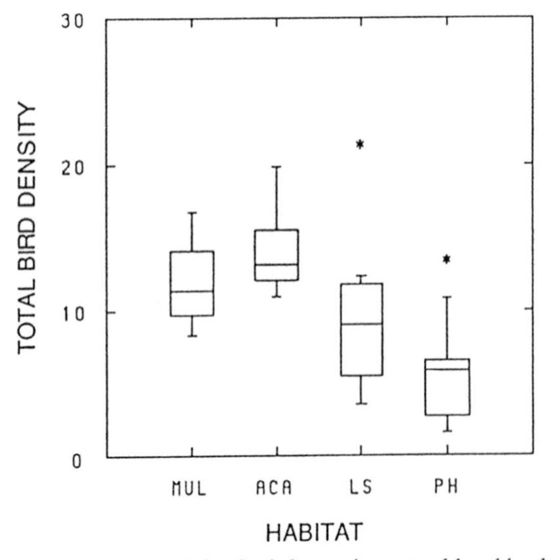

Fig. 5.17. Total bird density is lowest in protead heathlands, and increases to the highest levels in mulga and acacia-heath (difference between these two statistically non-significant).

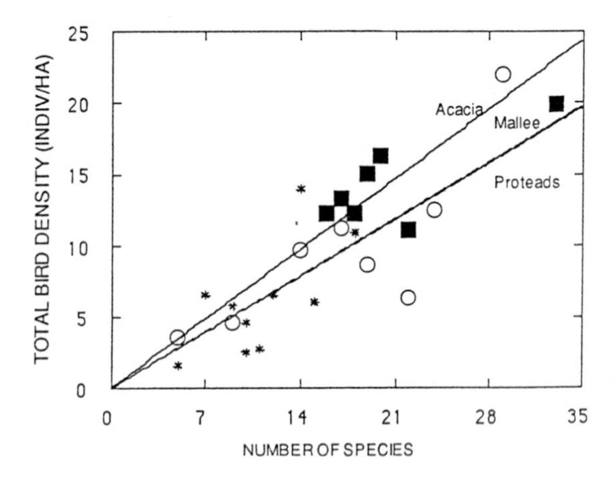

Fig. 5.18. Total bird density is plotted as a function of total species number. The relation is statistically significant for protead heathlands (open circles), but not for mulga (not shown), acacia-heath or mallee. A non-significant relation indicates density compensation, a significant relation implies strict resource partitioning and low niche overlap.

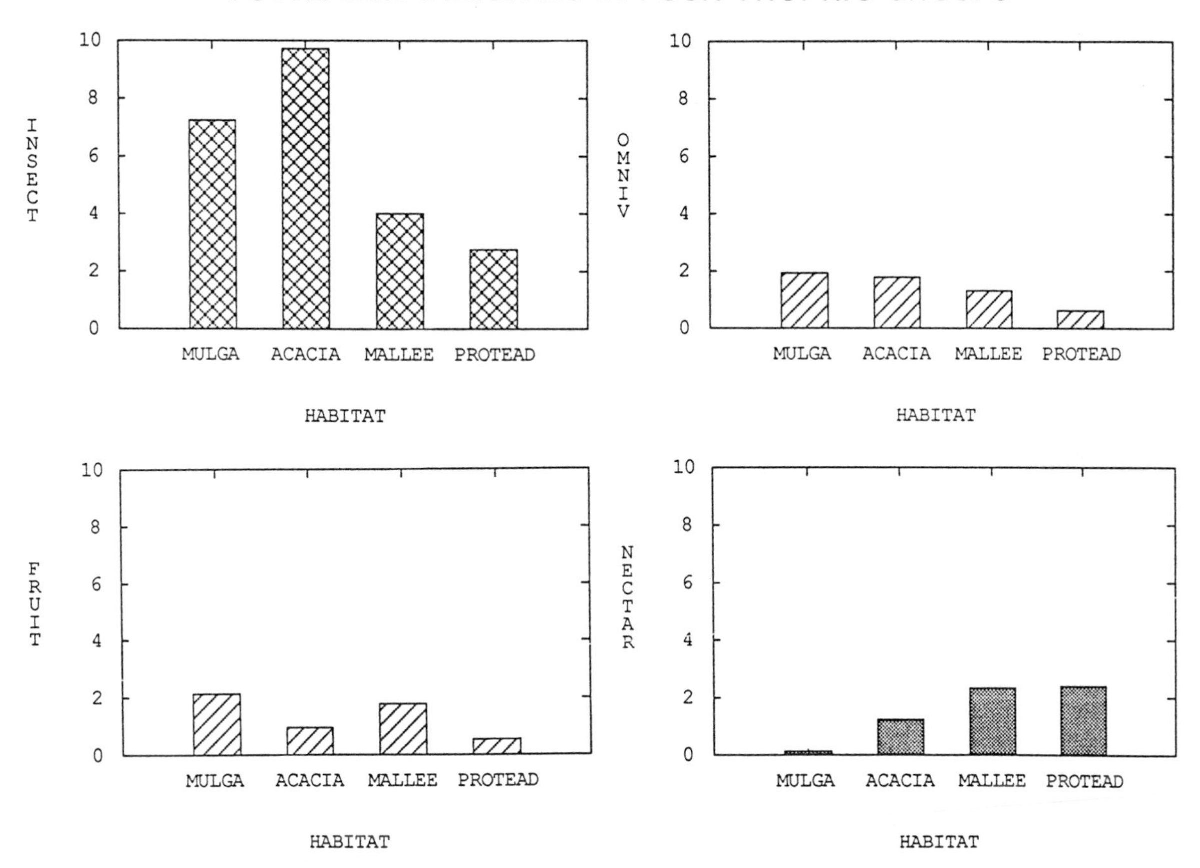

Fig. 5.19. Allocation of species to trophic guilds in four SW habitats. Insectivores are dominant in mulga and acacia-scrub, while nectarivores are common in protead heathlands and mallee.

be found in a wide range of PH habitats, from lower to taller, and less to more dense, vegetation; e) there is a significant distance effect, but γ-diversity increases relatively modestly with distance in the northeast, requiring 800–1000 km for 100% species turnover. Yet even very distant sites share many of the same species, which invade from nearby woodlands equally readily into PH sites even 2000 km apart.

Other heathlands

Mulga, acacia- and mallee-dominated heaths in southwestern Australia

In this final section, I ask how other types of heathlands, as well as mulga shrubland, differ from protead-dominated heathlands in terms of species composition and diversity components. I use sites in southwest Australia to answer the question. The similarities in vegetation structure among these habitats are shown in Fig. 5.11, where contour ellipses define 6 forest and woodland habitats of taller vegetation, and at similar vegetation height but increasing in vegetation density, mallee, acacia-scrub and protead heaths. Mulga is relatively taller and less dense than the heathlands. These four habitats can be ranked in terms of productivity from mulga (highest) through acacia-scrub and mallee to protead heathlands (lowest).

Bird α-diversity increases from protead-heaths (PH) through mallee-heaths (LS) to acacia-heaths

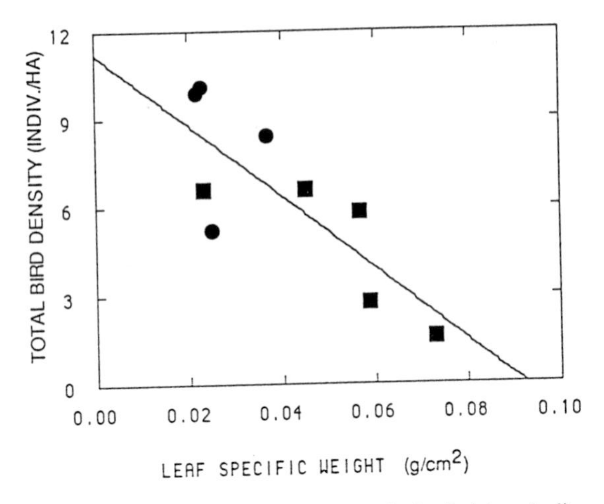

Fig. 5.20. Total bird density (ordinate; indiv./ha) is a declining function of mean leaf specific weight (LSW) of *Acacia* species at the site in protead heathlands, for both NE (filled circles) and SW (filled squares) sites. Leaf specific weight is an index of soil nutrient status, being highest in sites with the poorest soils.

(ACA), with mulga (MUL) supporting still higher species numbers (Fig. 5.12). Are these differences in α-diversity a simple consequence of increasing structural diversity from PH to mulga? This appears not to be so, for although LOGAREA predicts species numbers within PH habitats (Fig. 5.13), there is no significant relation within other heathlands or in mulga; mallee, acacia-scrub and mulga have increasing species numbers in vegetation of similar LOGAREA values.

Censuses in these four southwestern habitat types produce some 94 bird species; of these, 47 occur in mulga, 67 in ACA, 57 in PH and 52 in LS. Binomial models show that the distribution of species over sites differs among the different habitats (Fig. 5.14, observed to the left, expected to the right). There are many core species in the highly predictable mulga censuses, and few core species in the heaths. The predictable species mean a poor fit to the model in mulga, and the numerous rare or unpredictable species mean a poor fit in PH and ACA habitats. In LS, the fit with the binomial is good, with no too-common species and fewer rare species.

Using the joint criteria of most sites occupied and highest habitat-specific densities, 20 species are considered to be mulga-based, 10 ACA-based, 13 in LS and 10 species in PH (Table 5.2). A further 12 species are cosmopolitan with respect to habitat, and of the remaining 29, 20 occur at one site only. The ranking of species by habitat preferences suggests that the four habitats may be serially and parsimoniously arranged Mulga-ACA-LS-PH to reflect the species preferences (shown in Fig. 5.15). That is, species common or dominant in mulga become less dominant in sequence through ACA-LS-PH, and likewise in the other direction for species dominant in PH. Fig. 5.16 shows the attenuation of species, grouped according to habitat preference, in other habitat types within the series, thereby supporting the view that birds perceive these habitats as a gradient from mulga-ACA-LS-PH.

Bird densities and productivity

Total bird densities increase with species number through the heath habitats, but are lower in mulga even though more species are present (Fig. 5.17). Within the six mulga sites, there is no relation between species number and total bird density ($r = 0.554$, $p>0.05$), thereby indicating a density compensation among food-generalist bird species. There is a significant relation ($p<0.05$), however, between total density and species number in all three sorts of heathland (Fig. 5.18), with significantly higher density/species (slope 0.669) in ACA than in either PH or LS (slopes 0.563, 0.566 respectively). Correlation between species numbers and total density indicates strict resource partitioning among food specialists.

In terms of trophic relationships, the species composition changes drastically among the habitat types, with the mulga and acacia-heath supporting insectivores as the bulk of the bird community, and insectivores declining steeply in the mallee and protead-heaths (Fig. 5.19). Nectarivores are nearly absent from mulga, and increase along the habitat sequence to peak in protead-heaths; both fruit/seed eating and omnivory decline from mulga to PH habitats (see Fig. 5.19).

These patterns appear to reflect differences in the overall productivity of the sites, i.e. highest in mulga and lowest in PH. This relation can be confirmed indirectly by data on the leaf specific weights of *Acacia* species, which are available for a range of PH sites in both the southwest and northeast. Leaf specific weight is highest in sites of lowest soil nutrients, and sites with the highest values of leaf specific weights are those with the lowest bird densities ($r = 0.80$, $p<0.01$; Fig. 5.20). *Acacia* morphology thus serves as a reliable predictor of bird densities in heathland habitats.

Acknowledgments

Field work in Australia was conducted with financial assistance from the National Geographic Society, Washington, D.C., to which agency I am greatly indebted. Logistic help was gratefully accepted from the Departments of Zoology, University of Western Australia and University of Queensland. I am especially grateful to Dr. Jiro Kikkawa, University of Queensland, for his generous and invaluable advice on Queensland birds and habitats.

References

Cody, M.L. 1986. Diversity, rarity, and conservation in Mediterranean-climate regions. In: Soulé, M. (ed.), Conservation Biology. Sinauer Assoc., Sunderland, MA.

Cody, M.L. 1993. Bird diversity components within and among Australian habitats. Ch. 13 in: Ricklefs, R.E. & Schluter, D. (eds.), Species Diversity: Historical and Geographic Aspects. In press, Univ. Chicago Press, Chicago, Ill.

George, A.S., Hopkins, A.J.M. & Marchant, N.G. 1979. The heathlands of western Australia. In: Specht, R.L. (ed.), Heathlands and Related Shrublands, vol. 9A, Ecosystems of the World. Elsevier Scient. Publ. Co., Amsterdam, pp. 211–230.

Gilmore, A. 1985. The relation between vegetation structure and insectivorous bird densities. In: Keast, A., Recher, H.F., Ford, H. & Saunders, D. (eds.), Birds of Eucalypt Forests and Woodlands, Surrey Beatty & Sons Pty Ltd, Sydney, pp. 21–31.

Groves, R.H. 1981. Heathland soils and their fertility status. In: Specht, R.L. (ed.), Heathlands and Related Shrublands, vol. 9B, Ecosystems of the World, Elsevier Scient. Publ. Co., Amsterdam, pp. 143–150.

Keast, A. 1961. Bird speciation on the Australian continent. Bull. Mus. Comp. Zool. **123**: 305–495.

Keast, A. 1985. Bird community structure in southern forests and northern woodlands. In: Keast, A., Recher, H.F., Ford, H. & Saunders, D. (eds.), Birds of Eucalypt Forests and Woodlands. Surrey Beatty & Sons Pty Ltd, Sydney, pp. 97–116.

Kikkawa, J., Ingram, G.J. & Dwyer, P.D. 1979. The vertebrate fauna of Australian heathlands – an evolutionary perspective. In: Specht, R.L. (ed.), Heathlands and Related Shrublands, vol. 9A, Ecosystems of the World, Elsevier Scient. Publ. Co., Amsterdam, pp. 231–279.

Recher, H. 1985. A model of forest and woodland bird communities. In: Keast, A., Recher, H.F., Ford, H. & Saunders, D. (eds.), Birds of Eucalypt Forests and Woodlands. Surrey Beatty & Sons Pty Ltd, Sydney, pp. 129–135.

Sattler, P.S. (ed.) 1986. The Mulga Lands. Royal Society of Queensland, North Quay, Brisbane.

Specht, R.L. 1979. The sclerophyllous (heath) vegetation of Australia: the eastern and central states. In: Specht, R.L. (ed.), Heathlands and Related Shrublands, vol. 9A, Ecosystems of the World, Elsevier Scient. Publ. Co., Amsterdam, pp. 1–18.

Specht, R.L. 1981. Heathlands. In: R.H. Groves (ed.), Australian Vegetation. Cambridge Univ. Press, Cambridge, New York, Melbourne, pp. 253–275.

CHAPTER 6

Plant community structure in southwestern Australia and aspects of herbivory, seed dispersal and pollination

DAVID T. BELL
Department of Botany, The University of Western Australia, Nedlands, WA 6009, Australia

Key words: generalists, leaf-form spectra, mutualisms, plant-animal interactions, shrublands

Abstract. The plant community structure in southwestern Australia is influenced, in terms of plant/animal interaction, by 'generalists' as well as by the more-specialized forms of association between organisms. Leaf form spectra indicate a dominance of small, hard and spinous leaf forms in most plant communities. Small and hard leaves appear more likely to be responses predominantly to abiotic influences in the environment whereas leaf spines indicate a possible interaction with biotic influences. Documentation of food resources by western grey kangaroos and tammar wallabies indicate that these marsupials are versatile feeders and, thus, could not eliminate a particular species, but could have some long-term influence on leaf structure. Seed collection studies in the jarrah forest of Western Australia indicate that ants collect legume seeds with white arils and 'crinkled coatings', but vertebrate seed collectors are limited. Pollination interactions also indicate that pollinators often visit a number of plant species and 'generalist' flower morphologies may have advantages in the communities of southwestern Australia rich in plant species.

Introduction

The sclerophyllous shrublands, woodlands and forests of southwestern Australia occur in a landscape of limited topographic relief, depauperate soils and a Mediterranean-type climate with cool, wet winters and warm, dry summers (Beard, 1984). The vegetation shares a visual appearance common to the other five Mediterranean-type climate regions of the world. This superficial similarity of the vegetational response to common environmental stresses in these five Mediterranean-type climate regions, however, has changed, with more rigorous scientific scrutiny, to a wide range of reported differences. Barbour and Minnich (1990) note that the five regions generally have a vegetation of one layer, dominated by sclerophyllous, rigidly branching, interlocking shrubs; however, sometimes profound differences exist in species richness, life-form and leaf-form spectra, biomass profiles, biomass accumulation, phenology and response to fire. This chapter provides information on aspects of the plant/animal interaction of certain sclerophyllous shrublands (kwongan), woodlands (jarrah forest) and forests (karri forest) of southwest Western Australia and their influence on the structure of the plant communities in this mediterranean climate.

Leaf form – marsupial herbivory potentials

Collections of information on community leaf characteristics can reveal when environmental pressures may commonly be presented to the vegetation (Bell & Heddle, 1989). Information on leaf size, leaf consistency and leaf apex types have been collected for a range of Western Australian plant communities (Table 6.1). In Western Australia, two aridity gradients prevail, one west to east reflecting increasing continentality, and a second from south to north, reflecting a temperate to subtropical moisture trend. The plant community patterns in leaf size show a tendency for the communities in the far southwest to contain high proportions of species in the larger leaf sizes, which are gradually replaced by communities with greater proportions of species in the smaller leaf

M. Arianoutsou and R.H. Groves, Plant-Animal Interactions in Mediterranean-Type Ecosystems, 63–70, 1994.

Table 6.1. Leaf characteristics in Western Australian plant communities. Values are percentage of plant species categorized in each category from between 50–95 species in each location. AUG (Augusta), DON (Donnelly River), and DWL (Dwellingup) are sclerophyll forest communities. PER (Perup) is a woodland. GRN (Green Head) is a coastal duneslack and JUR (Jurien) and WAT (Watheroo) are kwongan shrublands. These communities represent a rough gradient of increasing aridity

Leaf characteristic	Plant community location						
	AUG	DON	DWL	PER	GRN	JUR	WAT
Leaf size							
Subleptophyll	10	6	4	6	5	14	19
Leptophyll	8	18	20	12	5	24	21
Nanophyll	26	31	25	42	58	42	30
Nano-microphyll	18	11	20	20	32	8	16
Microphyll	30	27	28	10	0	9	14
Micro-mesophyll	8	7	3	10	0	3	0
Leaf consistency							
Sclerophyll	26	–	18	30	4	45	75
Semi-sclerophyll	34	–	21	35	17	35	22
Mesophyll	10	–	15	0	11	2	0
Semi-succulent	0	–	0	2	24	2	0
Succulent	10	–	15	0	11	2	0
Leaf apices							
Acute	58	56	45	42	53	35	54
Spinose	34	28	46	48	37	48	36
Rounded	7	12	6	8	10	10	10
Cleft	0	4	3	2	0	7	0

size classes in more inland and more northern locations.

All Western Australian plant communities contain the hard, sclerophyll-type leaves and the more pliable class of semi-sclerophyllous leaves, but the kwongan shrublands on the inland of the Northern Sandplains show exceptionally large percentages of these leaf types. The community at Green Head (GRN) is somewhat atypical of the region, as a number of salt-succulent herbaceous species occur in this coastal dune and salt marsh region. The community leaf spectra accumulated on leaf apex types showed that the acute leaf apex was common; a finding not unexpected as most leaves tend to be longer than wide. Of greater interest regarding plant/animal interactions, however, was that all communities had from a third to a half of the species showing spinous leaf apices.

Interpretation of the trends in percentages can be construed as a response to marsupial grazing pressure. Small, hard and spine-tipped leaves make marsupial consumption less likely, especially as the native flora has very few species with branch or stem spines. In the semi-arid climates of southwestern Australia, however, small leaves dissipate energy more efficiently than larger leaves and sclerenchyma-strengthened leaves suffer less tissue damage under periods of drought than more mesic forms (Goble-Garratt *et al.*, 1981). These leaf characters could, therefore, be primarily responsive to the abiotic environment. The preponderance of leaf spines in the Western Australian shrublands and dry sclerophyll forests, appears likely, however, to be a community pattern influenced by biotic factors of the habitat.

Marsupial herbivory

Several studies have utilized epidermal remains in faecal pellets to provide direct evidence of marsupial herbivory in Western Australia; and a wide range of species have been documented as plant resources consumed by marsupial herbivores (Table 6.2). Many of these plant species have hard, spinous leaves, but preference has been slightly toward herbaceous perennials, tufted subshrubs and leguminous shrubs. Halford *et al.* (1984) recorded the food resources of the western

Table 6.2. Plant resource species documented for the Western Australlan marsupial herbivores, the western grey kangaroo, *Macropus fulginosus*, and the tammar wallaby, *Macropus eugenii*. Life form abbreviations: C – Climber; H – Herbaceous perennial; S – Shrub; SS – Sub-Shrub; T – Tree. Food resource species from Halford *et al.* (1984), Bell *et al.* (1987) and Bell, unpublished

Plant species	Life form	Food resources (×)	
		Kangaroo	Tammar
Acacia celastrifolia	S	×	
Acacia extensa	S	×	
Acacia rostellifera	S		×
Adenanthos cygnorum	S	×	
Allocasuarina spp.	T	×	
Anthocercis littorea	SS	×	
Asparagus asparagoides	C		×
Asphodelus fistulosus	H		×
Astroloma ciliatum	SS	×	
Baeckaea camphorasmae	S	×	
Bossiaea eriocarpa	S	×	
Bossiaea linophylla	S	×	×
Bossiaea ornata	S	×	
Burchardia multiflora	H	×	
Callitris preissii	T		×
Calothamnus sanguineus	S	×	
Conostylis setigera	H	×	
Clematis pubescens	C	×	×
Cyathochaeta avenacea	H		×
Dampiera lavandulacea	SS	×	
Daviesia decurrens	S	×	
Daviesia juncea	H	×	
Dianella revoluta	H	×	
Dryandra carduacea	S	×	
Eremophila glabra	SS		×
Gastrolobium bilobum	S		×
Gastrolobium calycinum	S	×	
Gastrolobium trilobium	S	×	
Glischrocaryon aureum	SS	×	
Gompholobium preissii	S	×	
Gyrostemon subnudus	S	×	
Hakea ambigua	S	×	
Hakea trifurcata	S	×	
Hakea undulata	S	×	
Hibbertia cunninghamii	S		×
Hibbertia racemosa	S	×	
Hypocalymma angustifolium	S		×
Jacksonia restioides	S	×	
Lasiopetalum molle	S	×	
Lasiopetalum oppositifolium	S		×
Lepidosperma scabrum	H	×	
Leucopogon capitellatus	S	×	×
Leucopogon pulchella	S	×	
Leucopogon verticillatus	S	×	
Lomandra effusa	H	×	
Lomandra hermaphrodita	H	×	
Lomandra preissii	H	×	
Loxocarya fasciculata	H	×	
Loxocarya flexuosa	H	×	
Macrozamia reidleii	S		×
Mesomalaena tetragona	H	×	
Mirbelia ramulosa	S	×	
Neurachne alopecuroidea	H	×	

Table 6.2. Continued.

Plant species	Life form	Food resources (×)	
		Kangaroo	Tammar
Opercularia vaginata	SS	×	
Persoonia longifolia	T	×	
Petrophile serruriae	S	×	
Phyllanthus calycinus	SS		×
Schoenus cyperacea	H		×
Solanum symonii	S		×
Stipa flavescens	H		×
Stylidium affine	H	×	
Stypandra imbricata	H	×	
Tetraria octandra	H	×	
Tetratheca confertifolia	SS	×	
Thomasia cognata	S		×
Trachyandra divericata	H		×
Tribonanthes uniflora	H	×	
Xanthorrhoea preissii	S	×	

grey kangaroo (*Macropus fuliginosus* Desm.) in a landscape of native woodland and pasture grasslands. The study showed that, even though animals were observed regularly feeding on the pastures at night, a considerable number of native woodland species appeared in their diet. *Conostylis setigera*, *Opercularia vaginata*, *Bossiaea eriocarpa* and the unresolved group of species of *Allocasuarina* were the favoured species of natives in the woodland. Following winter rains, the pasture grasses constituted the major fraction of the diet of the grey kangaroo, but native shrubs (especially the legumes) and herbaceous perennials were still found in considerable quantities in the diet.

A second direct study of herbivory by a population of the rare tammar wallaby (*Macropus eugenii* Desm.) on Garden Island, using both faecal analysis and stomach content analysis, also revealed the polyphagous nature of Western Australian marsupial herbivores (Bell *et al.*, 1987). Preference apparently was for the dominant shrub *Acacia rostellifera*, the introduced herbaceous species *Asphodelus fistulosus*, and the native grass *Stipa flavescens*. Animals fed on a range of species, both native to Western Australia and introduced, however. Results of exclosure studies demonstrated the potential for a marked impact of grazing on particularly favoured species and a preference for recent post-fire resprouts in the diet. The study also found that long-unburnt vegetation was important in providing cover for these small herbivores.

During the winter of 1991 a preliminary examination of dietary preferences of marsupial herbivores of the Perup Forest, a region of southern jarrah woodland that contains a remarkable number of rare mammals or those with restricted range, again revealed that marsupial herbivores cope with the diversity of potential resources by selecting a wide range of species. In this region both the western grey kangaroo and the tammar wallaby co-occur. Although these two marsupial species share some food resources and often overlap in home range, the versatility of their feeding abilities appears to allow co-habitation.

Western Australia has two genera of legumes, *Gastrolobium* and *Oxylobium*, which contain species that produce quantities of fluoroacetate poisons capable of killing introduced sheep and cattle (Gardner & Bennetts, 1956). The major marsupials, however, have evolved a tolerance to these toxins and their blood can often contain high levels of this allochemical (Main, 1968; Oliver *et al.*, 1977). The presence of *Gastrolobium calycinum* and *G. trilobium* in the diet of the Western Grey Kangaroo and *Gastrolobium bilobum* in the diet of the tammar wallaby confirms that these plant species are tolerated and in fact could be a major source of nitrogen. The presence of *Gastrolobium bilobum* in the diet of the tammar lends credence to one of the reasons for the appearance of this rare marsupial in the mainland Perup Forest area, an area also inhabited by introduced foxes. The tammar's tolerance to and retention of the contained allochemical could limit fox predation in this population. Thickets of this leguminous species also provide cover for the tammar; fire management in the Perup Forest is designed to favour this plant species (Christensen & Kimber, 1975).

Assumptions relating herbivores to the particular presence or absence of plant species in southwest Western Australia appear limited. Although the marsupial herbivores appear to favour some species and/or life forms, their general lack of concentration would mean that marsupials would not graze a species to extinction. Stresses other than herbivory are probably more important for plant species survival in this mediterranean climate habitat. Leaf spines, however, could be a community response to mild herbivore stress.

Seed dispersal

Australian plant communities contain numerous species with arillate seeds which are dispersed by ants (Buckley, 1982). Myrmecochorous species in Australia have been estimated to exceed 1500. In southern Africa ants facilitate escape from the parent in about 1000 species, but dispersal by ants has been documented in less than 300 species elsewhere (Buckley, 1982). Preliminary observations of seeds of most species in Western Australia reveals little of special morphological interest; most being black (probably a camouflage following fires), but with few appendages besides the arils or elaiosomes of particular *Acacia* species. In Western Australia reports of seed dispersal have been limited (Shea *et al.*, 1979; Majer, 1980) and speculation on the influence of animal dispersal agents on plant community even more limited.

Winter trials during 1989 and 1990 in a jarrah woodland site in the John Forrest National Park near Perth provided a selection of seeds under conditions where (1) invertebrates (presumably ants) only and (2) vertebrates only (probably only birds although possibly also mammals) could gain access. Seeds from a range of species, both native and introduced, were provided at the feeding stations for 48 h periods in both years and a number of replications occurred for most seed species (Table 6.3).

The invertebrates (probably solely ants) of the jarrah forest appear to rapidly collect fruits of a number of the *Acacia* species and all those collected in large numbers displayed white arils. Ants of the jarrah forests appear to be generalists, with single species of ants choosing arillate seed of a wide range of plant species. *Acacia cyclops* and *A. melanoxylon* were not collected, but these species display large, red-colored (bird-attracting?) arils and are not species native to the jarrah forest. *Acacia pendula* and *A. podalyrifolia*, two species of this genus without arils, were not favoured in collections by ants. *Gastrolobium bilobum*, an aril-containing legume, was also a favoured species. Another favoured species was *Gompholobium tomentosum*, a non-arillate legume species, but one with a 'crinkled coat' structure, which could provide a similar nutritional attraction to ants as arils. Birds and mammals do not appear to be important seed dispersal agents in Western Australian plant communities, although the jarrah forest contains a range of pigeons, quails and small doves. During periods of more severe shortage, however, the seed trays may receive more attention. Additionally, birds may avoid the feeding trays and here seed collection by these animals is, thus, underrepresented by this technique (Kerley, 1991).

Dispersal of seeds away from the influence of the parent can be important (Westoby *et al.*, 1982). In Western Australia, however, it is probably the ability of seeds to germinate in years following fires and at appropriately moist periods of the year, which are the more important attributes for species survival (Bellairs & Bell, 1990; Bell & Bellairs, 1992). In the ant/legume association, however, one sees a positive plant/animal interaction capable of influencing the structure of plant associations in these southwestern Australian communities.

Table 6.3. Percentages of available seed collected by invertebrates or vertebrates of the jarrah forest of the John Forrest National Park near Perth, Western Australia during one 48 h period in each of the winters of 1990 and 1991. WA = white-colored aril; RA = red-colored aril

Plant species	Percentage 'removal' of seed		
	Reps	Invertebrates	Vertebrates
Acacia baileyana-WA	3	15 ± 16	0 ± 0
Acacia cyclops-RA	2	4 ± 5	7 ± 4
Acacia decurrens-WA	2	26 ± 20	6 ± 8
Acacia extensa-WA	7	33 ± 17	5 ± 8
Acacia glauceptera-WA	3	13 ± 17	6 ± 6
Acacia orridula-WA	8	36 ± 38	0 ± 0
Acacia lasiocalyx-WA	3	49 ± 25	3 ± 4
Acacia lasiocarpa-WA	2	52 ± 6	12 ± 17
Acacia melanoxylon-Ra	7	13 ± 24	0 ± 0
Acacia pendula	6	3 ± 3	1 ± 2
Acacia podalyrifolia	6	11 ± 11	1 ± 2
Acacia pulchella-WA	9	23 ± 24	10 ± 22
Acacia redolens-WA	5	27 ± 32	5 ± 5
Acacia steedmanii-WA	5	58 ± 42	5 ± 7
Acacia vestita-WA	2	44 ± 42	0 ± 0
Asphodelus fistulosus	9	7 ± 11	1 ± 1
Bauhinia purpurea	21	1 ± 1	0 ± 0
Bossiaea ornata-WA	8	13 ± 13	6 ± 7
Burchardia umbellata	7	3 ± 7	4 ± 9
Citrullus lanatus	11	3 ± 7	0 ± 0
Cucumis melo var *cantalupensis*	14	7 ± 8	4 ± 8
Eucalyptus erythrocorys	6	12 ± 24	0 ± 0
Eucalyptus marginata	8	6 ± 10	1 ± 2
Eucalyptus wandoo	5	4 ± 9	5 ± 7
Gastrolobium bilobum-WA	7	58 ± 39	5 ± 9
Gompholobium tomentosum	9	33 ± 29	7 ± 9
Hakea lissocarpha	7	16 ± 18	1 ± 3
Hypocalymma angustifolium	6	17 ± 26	4 ± 6
Hypocalymma robustum	3	3 ± 4	4 ± 4
Phoenix canariensis	4	0 ± 0	0 ± 0
Phyllanthus calycinus	10	16 ± 19	3 ± 5
Pittosporum phyllaeroides	7	7 ± 13	3 ± 4
Ricinus communis	20	4 ± 10	3 ± 9
Sollya heterophylla	4	30 ± 27	6 ± 9
Stirlingia latifolia	13	7 ± 10	1 ± 1
Trachymene octandra	4	15 ± 12	4 ± 3
Trymalium ledifolium	7	12 ± 22	5 ± 10
Trymalium spatulatum	2	0 ± 0	0 ± 0

Pollination

Specialized interactions between particular insects and particular plants (e.g. the pseudocopulatory interaction between thinnid wasps and particular terrestrial orchids) have highlighted past research on the plant/animal interactions associated with pollination in Western Australia. Visitor constancy followed by evolutionary fitting of flower morphology and specialization of the pollinator is a common theme in many pollination interactions (Stiles, 1981; Waser, 1987; Galen *et al.* 1987). In Australia, however, almost all bees are solitary (Armstrong, 1979) and the paucity of endemic social bees may explain the prominence of other pollinators in Australian plant communities (Ford, 1985a,b). Australian native bees tend to be promiscuous flower visitors and, therefore, there may be some advantage for plant species to maintain 'generalized' flower forms. Recent studies of the introduced European honey bee visiting kwongan communities in Western Australia show that these insects tend to visit single species on particular pollen collection forays, but they are also versatile feeders, visiting a wide range of species (Van der Moezel *et al.*, 1987; Wills *et al.*, 1990). Of the 125 flowers visited by the European honeybee in the kwongan region of the Beekeepers Reserve in the Northern Sandplains of Western Australia, 74% were of the 'typical' insect pollination types, 20% were more commonly associated with a vertebrate pollination syndrome and 6% were best adapted to wind pollination (Wills, 1989). The most common resources for the honeybee in these Western Australian communities were woody perennial species that flowered for long periods, had a widespread or locally abundant distribution and most often were species which require establishment from seed following fires. The ability of the introduced *Apis mellifera* to utilize profitable resources regardless of the apparent pollination syndromes of the plants has been observed even where the plants have co-evolved with the insect (Adams *et al.*, 1978). In the highly diverse plant assemblages of Western Australia, attraction of numerous pollinators may have distinct advantages. MacArthur and Pianka (1966) and Levin (1978) argue that the diet of animal foragers should broaden as the environment becomes more variable. This may be the case in the Western Australian kwongan when bees are faced with such a diverse assortment of possible flower resources. Many plant species of the kwongan shrublands appear to attract a wide range of potential pollinators and 'generalist' flower shapes seem more common than specialized types that are likely related to particular plant-animal pollination interactions.

Conclusions

In the past it was fashionable to emphasize the tight inter-relationship between particular plant species and a particular animal; these two organisms associated in some 'reciprocal co-evolutionary battle' waxing to and fro over millennia. Western Australian data appear to indicate that most animals are generalists and tend to present occasional pressure to individuals, but rarely is there a population pressure which could be related to the favouring of particular members of the plant population with particular attributes. Marsupial grazers are versatile feeders, probably having some, but only some, effect on the flora. Favorable dispersal by ants has lead to a number of plant species developing anatomical structures to enhance movement away from the parent, but physiological responses to environmental stimuli probably are more important to the long-term survival of populations. Pollination interactions in the highly diverse plant communities of Western Australia indicate some advantage in maintaining a 'general' flower morphology but further research is needed. It may be in the sclerophyllous Mediterranean-type plant communities that 'generalist feeders' are most common and these animals have only limited effect on the majority of plant species sharing these seasonally severe climates.

Acknowledgements

I am indebted to my 1990 and 1991 third year classes in Plant/Animal Interactions for assistance in the documentation of marsupial herbivory in the Perup Forest and seed collection patterns in the jarrah forest. Funding for the research on honey bee/plant interactions was from the Honey Research Board of the Australian Department of Primary Industry. The manuscript was prepared while on sabbatical leave at Coventry Polytechnic, whose cooperation and assistance was invaluable.

References

Adams, R.J., Manville, G.C. & McAndrews, J.H. 1978. Comparisons of pollen collected by a honeybee colony with a modern wind-dispersal pollen assemblage. Can. Field Nat. 92: 359–368.

Armstrong, J.A. 1979. Biotic pollination mechanisms in the Australian flora – a review. N.Z. J. Bot. 17: 467–508.

Barbour, M.G. & Minnich, R.A. 1990. The myth of chaparral convergence. Isr. J. Bot. 39: 453–463.

Beard, J.S. 1984. Biogeographlv of the kwongan. In: Pate, J.S. & Beard, J.S. (eds). Kwongan: plant life of the sandplain. University of Western Australia Press, Nedlands, W. Australia, pp. 1–26.

Bell, D.T. & Bellairs, S.M.. 1992. Temperature effects on germination of selected Australian native species used in rehabilitation of bauxite mining disturbance in Western Australia. Seed Sci. Technol. 20: 47–55.

Bell, D.T. & Heddle, E.M. 1989. Floristic, morphologic and vegetational diversity. In: Dell, B., Havel, J.J. & Malajczuk, N. (eds.). The jarrah forest – a complex mediterranean ecosystem. Kluwer Academic Publishers, Dordrecht, The Netherlands, pp. 53–66.

Bell, D.T., Mordoundt, J.A. & Loneragan, W.A. 1987. Grazing pressure by the tammar (*Macropus eugenii* Desm.) on the vegetation of Garden Island, Western Australia, and the potential impact on food reserves of a controlled burning regime. J. Roy. Soc. West. Aust. 69: 89–94.

Bellairs, S.M. & Bell, D.T. 1990. Temperature effects on the seed germination of ten kwongan species from Eneabba, Western Australia. Aust. J. Bot. 38: 451–458.

Buckley, R.C. (ed). 1982. Ant-plant interactions in Australia. Dr W. Junk, The Hague.

Christensen, P. & Kimber, P. 1975. Effect of prescribed burning on the flora and fauna of south-west Australian forests. Proc. Ecol. Soc. Aust. 9: 85–106.

Ford, H.A. 1985a. Nectar-feeding birds and bird pollination: why are they so prevalent in Australia yet absent from Europe? Proc. Ecol. Soc. Aust. 14: 153–158.

Ford, H.A. 1985b. Nectarivory and pollination by birds in southern Australia and Europe. Oikos 44: 127–131.

Galen, C., Zimmer, K.A. & Newport, M.E. 1987. Pollination in floral scent morphs of *Polemonium viscosum*: a mechanism for disruptive selection on flower size. Evolution 41: 599–606.

Gardner, C.A. & Bennetts, H.W. 1956. The toxic plants of Western Australia. W.A. Newspapers Ltd., Perth.

Gobble-Garratt, E.M., Bell, D.T. & Loneragan, W.A. 1981. Floristic and leaf structure patterns along a shallow elevational gradient. Aust. J. Bot. 29: 329–347.

Halford, D.A., Bell, D.T. & Loneragan, W.A. 1984. Diet of the western grey kangaroo (*Macropus fuliginosus* Desm.) in a mixed pasture-woodland habitat of Western Australia. J. Roy. Soc. West. Aust. 66: 119–128.

Kerley, G.I.H. 1991. Seed removal by rodents, birds and ants in the semi-arid karoo, South Africa. J. Arid Environ. 20: 63–69.

Levin, D.A. 1978. Pollinator behaviour and the breeding structure of plant populations. In: Richards, A.J. ed. The pollination of flowers by insects. Academic Press, New York, pp. 133–150.

MacArthur, R.H. & Pianka, E.A. 1966. On optimal use of a patchy environment. Am. Nat. 100: 603–609.

Main, A.R. 1968. Physiology in the management of kangaroos and wallabies. Proc. Ecol. Soc. Aust. 3: 96–105.

Majer, J.D. 1980. The influence of ants on broadcast and naturally spread seeds in rehabilitation bauxite mined areas. Reclam. Rev. 3: 3–9.

Oliver, A.J., King, D.R. & Mead, R.J. 1977. The evolution of resistance to fluoroacetate intoxication in mammals. Search 8: 130–132.

Shea, S.R., McCormick, J. & Portlock, C.C. 1979. The effect of fires on regeneration of leguminous species in the northern jarrah (*Eucalyptus marginata* Sm.) forest of Western Australia. Aust. J. Ecol. 4: 195–205.

Stiles, F.G. 1981. Geographic aspects of bird-flower coevolution with particular reference to central America. Ann. Missouri Bot. Gard. 68: 232–351.

van der Moezel, P.G., Delfs, J.C., Pate, J.S., Loneragan, W.A. & Bell, D.T. 1987. Pollen selection by *Apis mellifera* in shrublands of the Northern Sandplains of Western Australia. J. Apicult. Res. 26: 224–232.

Waser, N.M. 1987. Spatial genetic heterogeneity in a population of the montane perennial plant *Delphinium nelsonii*. Heredity 58: 249–256.

Westoby, M., Rice, B., Shelley, J.M., Haig, D. & Kohen, J.L. 1982. Plant's use of ants for dispersal at West Head, New South Wales. In: Buckley, R.C. (ed). Ant-plant interactions in Australia. Dr. W. Junk, The Hague, pp. 75–88.

Wills, R.T. 1989. Management of the flora utilised by the European honey bee in kwongan of the Northern Sandplain of Western Australia. Ph.D. Thesis, University of Western Australia, Nedlands, Australia.

Wills, R.T., Lyons, M.N. & Bell, D.T. 1990. The European honey bee in Western Australian kwongan: foraging preference and some implications for management. Proc. Ecol. Soc. Aust. 16: 167–176.

PART THREE

Triangular relationships

CHAPTER 7

Resource webs in Mediterranean-type climates

H.A. MOONEY[1] and R.J. HOBBS[2]
[1]*Department of Biological Sciences, Stanford University, Stanford, CA 94305, USA*
[2]*CSIRO Division of Wildlife and Ecology, LMB 4, PO, Midland, WA 6056, Australia*

Key words: environment-mediated interactions, gophers, serpentine grassland, ants, disturbance

Abstract. Whilst much attention has been paid to direct biotic interactions between plants and animals, this chapter shows that important interactions may be mediated by way of plant or animal effects on the environment. We term these complex interactions 'resource webs', in which each trophic level can influence the availability of basic resources, which in turn can affect other organisms. The concept is illustrated with a number of examples drawn primarily from studies of the effects of gopher and ant activities in annual grasslands found on serpentine in California and other regions of Mediterranean-type climate. The existence of environment-mediated interactions indicates that plant-animal relationships are important not only at the population level but also at the level of ecosystems as well.

Introduction

Although there has been considerable attention given to the study of plant-animal interactions (e.g., pollination biology, dispersal ecology, herbivore-plant interactions – see other chapters in this volume) less attention has been given to plant-animal interactions mediated through effects of either plants or animals on the environment. Animals, for example, can influence the structure of plant communities directly through foraging activites as well as either directly, or indirectly, through habitat alteration (Naiman, 1988). Many of these effects in turn alter the structure of animal populations (Fig. 7.1). In this chapter we consider the effects of animals on habitats and consequently on plant and animal populations in Mediterranean-type climates. We term these complex interactions among organisms and their resource base 'resource webs'.

Gopher-soil resource-plant interactions

We first examine the action of fossorial animals in Mediterranean-climate systems, with most emphasis on the Californian pocket gopher, *Thomomys bottae*. All Mediterranean-climate regions, with the exception of southern Australia, have convergent fossorial herbivorous rodents (Fig. 7.2) (Nevo, 1979) which have comparable impacts on the environment. Their role in structuring the landscape has been discussed recently at both the local (Huntly & Inouye, 1988; Huntly, 1991) and regional scales (Mielke, 1977).

California example – Jasper Ridge

Jasper Ridge, lies in San Mateo County, south of San Francisco, California. It is located within a Mediterranean-type climate with little or no rainfall between May and September. Annual precipitation is about 480 mm. The plant communities found on the Ridge include chaparral, oak woodland, and grassland. Much of the grassland occurs on soils derived from serpentine. This grassland has been the subject of considerable investigation in recent years, with a particular emphasis on the role of small mammals and ants on structuring the vegetation.

The vegetation of the grassland is very patchy

M. Arianoutsou and R.H. Groves, Plant-Animal Interactions in Mediterranean-Type Ecosystems, 73–81, 1994.

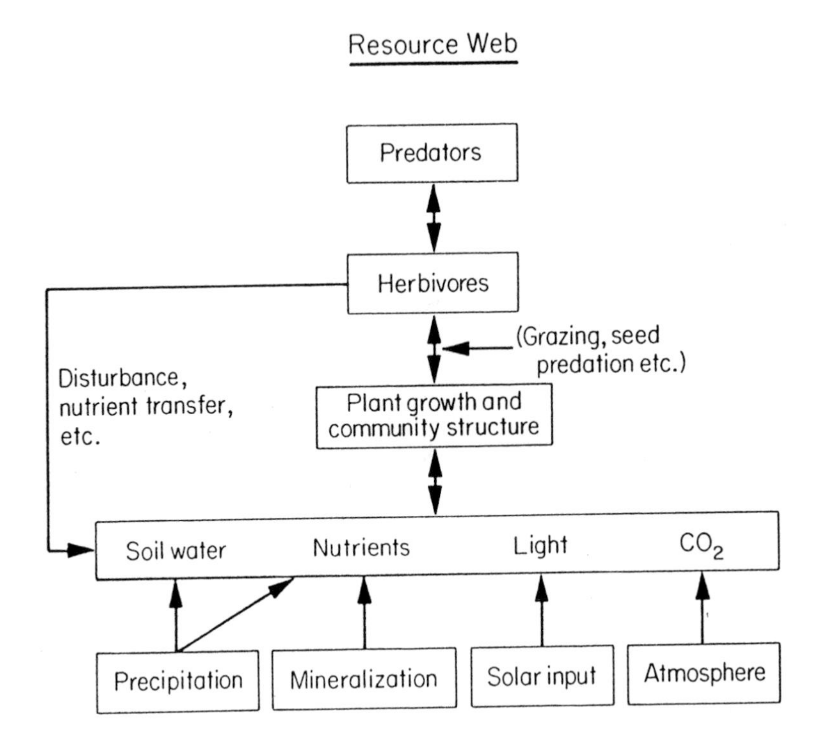

Fig. 7.1. Resource web structure.

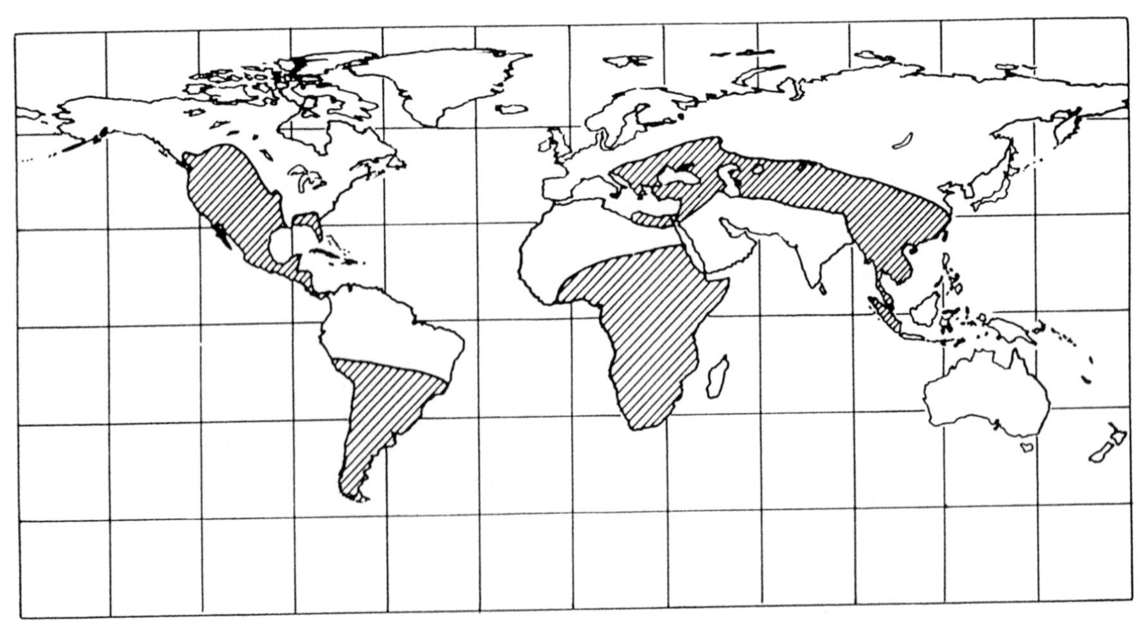

Fig. 7.2. Distribution of herbivorous fossorial rodents (from Nevo, 1979).

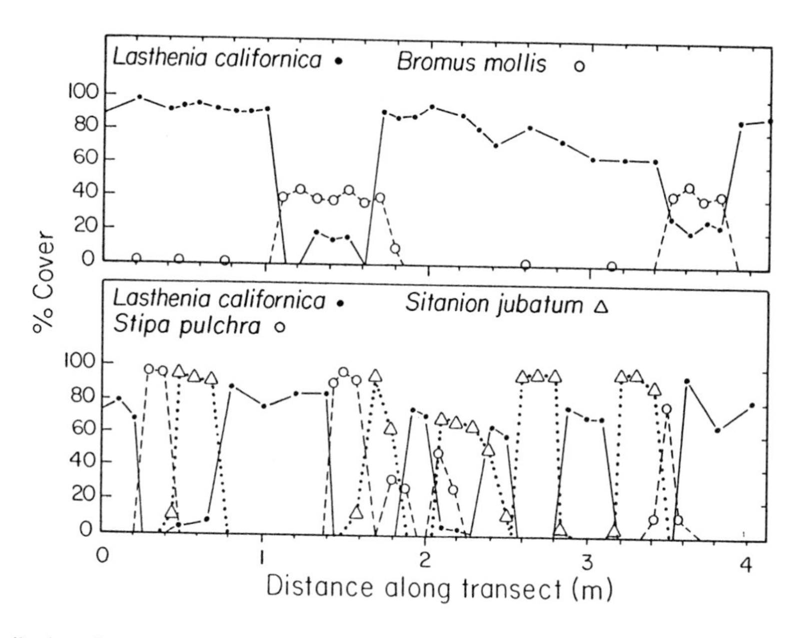

Fig. 7.3. Patchy distribution of plants in a gopher-disturbed habitat. Percentage cover of selected species along two 4 m transects in serpentine grassland at Jasper Ridge, California (from Hobbs & Mooney, 1985).

(Fig 7.3). This patchiness corresponds with the high degree of disturbance caused by the pocket gopher, which is considerable. In some years pocket gophers can turn over 40% of the soil surface with an average year being about 20% (Hobbs & Mooney, 1991). With such a high rate of activity there are very few microsites that do not experience massive disturbance over a period of several years (Fig. 7.4).

The gophers are most intensively active during the period of high soil moisture availability – autumn, winter and spring in the Californian Mediterranean-climate region (Miller, 1948; Cox *et al.*, in press). In desert sites, in contrast, the activity of *T. bottae* is regulated more by food supply and mate-searching activity than by rainfall (Bandoli, 1981).

Microsite effect

The digging activity of the gopher alters the microsite considerably. Subsurface soils are brought to the surface, thereby covering the existing vegetation. Consequently, soil surface temperatures increase because of the reduction in vegetative cover, nitrate availability increases because of enhanced decomposition, soil moisture decreases at the surface because of the increased radiation and temperature but increases at depth because of the reduction in the cover of transpiring plants (S.P. Hamburg, personal communication). Furthermore, sub-surface soils are of lower nutritional quality, they have lower contents of total nitrogen and phosphorus and higher levels of manganese (Koide *et al.*, 1987). They also have a lower potential for mycorrhizal infection which is important in this system in which over 90% of the plant species are infected with vesicular-arbuscular mycorrhizae (Chiariello *et al.*, 1982; Koide & Mooney, 1987).

Thus, a number of changes would appear detrimental to plant establishment and growth, whereas others appear beneficial. In pot experiments plants of *Plantago erecta* and *Bromus mollis* did more poorly when grown on mound versus intermound soils. Growth was reduced in part, but not entirely, because of reduced total N and P levels (Koide *et al.*, 1987).

In the field, contrary to predictions from pot experiments, plants actually do better on gopher mound soils (Fig. 7.5). This effect apparently is because of the many fewer individuals that occur on the mounds, and hence their reduced competition for the limited resources. In turn, the reduc-

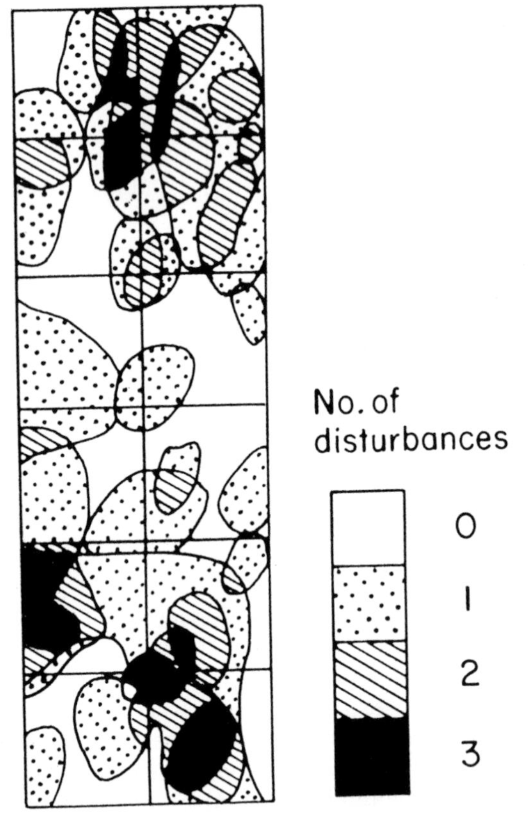

Fig. 7.4. Gopher disturbance history of a 1 × 3 m grid in an annual grassland in California over a 6-year period (from Hobbs & Mooney, 1991).

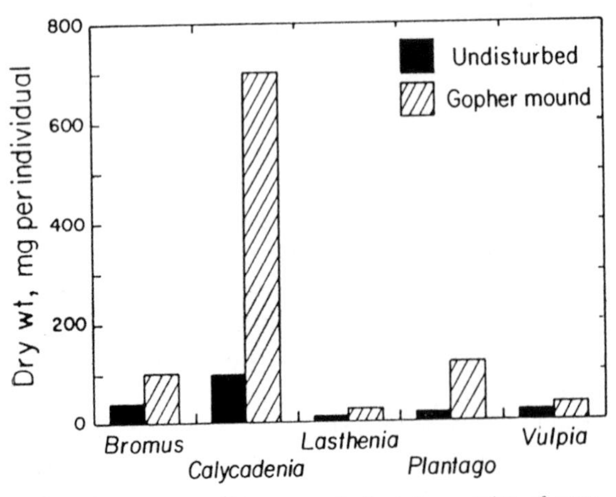

Fig. 7.5. Individual above-ground plant dry weights of annuals growing on and off gopher mounds in a California grassland (from Hobbs & Mooney, 1985).

tion in the numbers of individuals found on the gopher mounds arises in large part because of the limited dispersability of the serpentine plant species (Hobbs & Mooney, 1985), but could also be because the gopher mound surface is an unfavourable site for germination, a possibility that has not been investigated.

Gopher activity is heavily concentrated in the serpentine grassland although the soil is shallow and rocky. There are fewer gophers in adjacent grasslands that are developed on deep soils derived from greenstone which is comparatively easy to cultivate. This preference for serpentine by the gophers, in spite of the greater energy that must be expended in foraging, is apparently because of the presence of their preferred food plant at this site, viz. the geophyte *Brodiaea*, which is absent from greenstone (Proctor & Whitten, 1971).

Dispersal effects

The physical and chemical alteration of plant microsites no doubt plays a role in the community development on the gopher mound surfaces. Plant dispersal timing in relation to the timing of gopher mound production also is important, however, and may predominate the interactions. Hobbs & Mooney (1985) showed that the composition of the plants growing on gopher mounds depended on the time of formation of the mounds since the composition of the seed rain varied through time (Fig. 7.6).

Feedbacks

The gophers thus drive the patchy nature of the soil resources, such as nutrients and water, as well as the patchy cover of the vegetation. This patchiness, in turn, generates further patchiness since gophers prefer to work over sites with higher soil nutrient levels and consequently higher plant productivity (Hobbs *et al.*, 1988). There are also inter-annual controls on the gopher-environment-plant interactions. In years of differing rainfall amounts, the amount of gopher activity varies as does the flora available for colonization (Hobbs & Mooney, 1991).

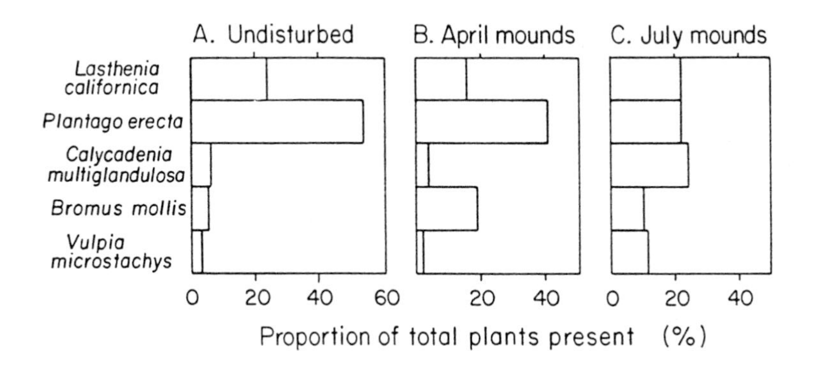

Fig. 7.6. Composition of plants on undisturbed and gopher-disturbed sites in an annual California grassland (from Hobbs & Mooney, 1985).

Gopher-soil resource-plant-insect interactions

Larvae of the butterfly *Euphydras editha* on Jasper Ridge suffer over 90% mortality because of loss of its principal food plant, the annual *Plantago erecta*, as the latter reaches the end of its life cycle and dries up as the annual drought commences (Singer & Ehrlich, 1979). Individuals of *Plantago* have a longer life by several weeks when growing on gopher mounds. This greater longevity is evidently because of the reduced competition that they experience in these microsites. Larval success of *Euphydras* is greatly enhanced when they are growing on plants occurring on gopher mounds.

So the success of the butterfly is directly related to the effect of gophers on providing a microsite on which its principal food plant is especially successful. Huntly and Inouye (1988) describe another instance where gophers affect insect populations through habitat alteration. Grasshoppers at their Minnesota study site increase in abundance where gophers are present because of the more favorable open soil microsites provided for egg deposition and development.

On convergence and co-evolution

Nevo (1979) has discussed the remarkable convergence in morphology and behaviour, and McNab (1966) the metabolism of fossorial animals throughout the world. In all Mediterranean climate regions, with the exception of southern Australia where they evidently are resource-limited (Cox *et al.*, in press), there are representative subterranean herbivorous mammals.

There have been a number of recent studies on the behavior and habitat impact of mole rats in South Africa. Although the mammal fauna of the fynbos is not particularly important in terms of biomass, mole rats may play a particularly important role in the overall ecosystem dynamics. In the Cape Region where mole rat densities can be high, Davies and Jarvis (1986) compared the behaviour of *Bathyergus suillus* and *Cryptomys hottentotus*. These animals have differing burrow depths, because of the soil type they inhabit, and they differ in their food sources, *Bathyergus* being a generalist, and *Cryptomys* feeding principally on geophytes, which they store. *Bathyergus* occupies sandy soils and has its burrows at some depth, even exceeding the depth of the roots of the many species upon which it feeds. It is active year-round with the greatest activity of mound building at the end of winter, when the soil is still moist. *Cryptomys* with its shallower burrows has its greatest activity in spring, coinciding with the breeding season. These results, along with those of other studies, indicate that soil type and resource use and availability, interact with overall climatic patterns to determine the seasonal activity cycles and foraging patterns of fossorial animals. For example, the Southwest African mole rat, *Cryptomys damarensis* '. . . are very local . . . , seldom going far from certain spots, doubtless where certain bulbs or roots are always to be found, but not necessarily, as one would expect, where soil is loose or sandy, very often remaining in the stoniest and hardest ground'

(Shortridge, 1934). This behaviour is comparable to that of the Californian pocket gopher which, as noted earlier, will forage on rocky soils in search of a favoured geophyte and avoid nearby sandy soils supporting a different flora that lacks geophytes.

Those mole rats that eat and store geophytes may have co-evolved with them. The debudding of geophytes to prevent them from sprouting in stores has been described for *Georychus campensis* (Shortridge, 1934) and *Cryptomys hottentotus* (Davies & Jarvis, 1986) in South Africa, as well as the Mediterranean mole rat, *Spalax ehrenbergi* (Galil, 1967). Lovegrove and Jarvis (1986) proposed that mole rats serve as a dispersal agent for the iris, *Micranthus junceus*. This plant is highly palatable to mole rats, and is stored by them in caches. The morphology of the iris is such, however, that the probability of a mole rat losing a viable portion of the plant while processing the plant is quite high – thus they are propagated in the process of being devoured! Galil (1967) similarly describes the role of mole rats in the dispersal of *Oxalis* and Cook (1939), the dispersal of thistle rhizomes by the California gopher.

It may be that gophers promote the development of favoured food plants indirectly through effects on the environment. As noted above, gopher activity can stimulate the release of nitrate. A number of plant species are stimulated to germinate in the presence of nitrate, such as the wild radish which the gophers eat (personal observations).

In Chile, the remarkable fossorial rodent, *Spalacopus cyanus*, has been described by Reig (1970). This animal may form almost continuous colonies for distances of 50 km in the central coastal zone. These colonial animals are specialists on the geophyte, *Leucoryne ixiodes*, and apparently move from site to site as this food resource is locally depleted. These animals have both a strong direct, as well as an indirect, effect on plant populations (Contreras & Gutiérrez, 1991).

Ants – direct and indirect impacts on plants

Ants can directly and indirectly affect plant populations. The activity around ant nests may de-

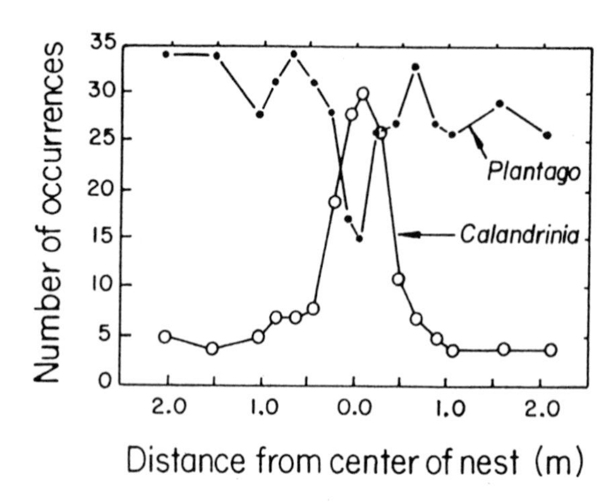

Fig. 7.7. Species patterning associated with ant nests in a California grassland (N. Chiariello, pers. comm.).

crease soil bulk density and generally increase nutrient levels, particularly phosphate (Buckley, 1982; Beattie, 1989). The patchy distribution of nests thus produce a patchy pattern of soil properties representing, for example, hot spots of phosphorus. N. Chiariello (personal communication) has found that California annual plants in serpentine grasslands may be negatively or positively associated with ant nests (Fig. 7.7). The positive associations, such as for *Calandrinia*, may represent a preference for the localized soil properties at the nest site, since these seeds are not collected by the resident ant seed-harvester, *Veromessor andrei*. The localized success of *Calandrinia* may be related to its lack of mycorrhizal association and the possible compensating effect of the enriched phosphorus near the ant nests. On the other hand, *Plantago erecta*, which is harvested by the ant, and which has low dispersability (Hobbs, 1985) is negatively associated with ant mounds. Thus, in this system, both direct and indirect effects of ants play a role in structuring the plant community.

On convergence and co-evolution of ants

Milewski and Bond (1982) provide a test of whether there has been convergence in ant-plant relations in Mediterranean-climate systems. They found that about one quarter of the plant species in sclerophyllous woodlands produced myrmecochorous, diaspores with elaisomes, or specialized ant-food bodies, in infertile sites of carefully matched Mediterranean-climate regions of South Africa and Australia. In sites with higher fertility in South Africa this amount dropped to 13%. Ninety percent of all of the myrmecochorous species that exist are restricted to South Africa and Australia (Buckley, 1982), areas noted for their low soil fertility. In contrast, in the higher fertility sclerophyll vegetation of California only one strict myrmecochore has been described, *Dendromecon rigida* (Berg, 1966). The greater abundance of seed-eating rodents in the California chaparral may also play a role in the lack of myrmecochory found there (cf. Milewski & Bond, 1982).

In spite of the richness of both the floras and ant faunas of South Africa and Australian regions of Mediterranean climate, Milewski and Bond (1982) found no examples of species-specific plant-ant interactions in the either Australia or South Africa.

Animal-environment-plant-insect interactions

Landsberg *et al.* (1990) have recently provided a striking example of the impact of animals on the environment that in turn affects plant properties and hence insect populations. In the Australian woodlands that have been converted to pasture, woodland remnants are now suffering dieback. They found that livestock that congregated near the trees enriched the soil nutrient level which, in turn, increased the leaf tissue concentration of nitrogen and phosphorus. This increased leaf nutrition increased the population sizes and larval growth rates of defoliating beetles.

Similar effects of congregating livestock can be seen in the biology of thistles and hence their insect herbivores. Thistles now have a widespread distribution throughout the temperate region and may be particularly weedy in Mediterranean-climate regions. The milk thistle, *Silybum marianum*, is of European origin but is now found as a weed in all of the Mediterranean-climate regions: Chile (Matthei, 1963), South Africa, California (Goeden, 1971), and Australia (Michael, 1968). This weed does particularly well in Mediterranean climates where the lack of late season competitors favours it (Michael, 1968).

Goeden (1971) has noted that the milk thistle has an affinity for fertile soils, such as is produced where livestock congregate, and that the plants can accumulate toxic levels of nitrate in these habitats. Michael (1968) showed that the milk thistle grows in microsites with twice the soil nitrogen level as in adjacent sites where it did not occur, and Austin *et al.* (1985) found that the milk thistle dominated five other thistle species at the highest nutrient levels they utilized in their experiments.

It would appear that livestock produce microsites that are particularly conducive to the development of populations of milk thistle. These thistles in turn support a diverse and specialized fauna in their native habitats and, to a lesser degree, in areas where it is a weed. For example, none of the insects found on it in Southern France were also found on it in Southern California, where it is a weed, although there were a great number of localized species associated with it in California, mostly generalists, that were not 'intimately attuned ontogenetically to the morphology or phenology of this plant' (Goeden, 1971).

Conclusions

In this chapter we have illustrated that plant-animal interactions include more complex relationships than those traditionally considered. Whilst herbivory, pollination and dispersal are undoubtedly important processes (see other chapters), there is another set of processes whereby plants and animals interact by modifying resource supplies. As a modification of the classical idea of food webs, which are more or less linear in their organisation (Pimm, 1982), we have introduced the idea of a resource web in which each trophic level can influence the availability of the basic resources of light, nutrients and water (Fig. 7.1).

The examples we have given illustrate a number of situations where animal activity (such as disturbance or nutrient transfer) has locally altered the environment and thus modified plant growth or plant community structure or composition. Examples can also be cited where plants modify the environment for animals, which then have further ecosystem effects. For instance, in situations where the shrub *Baccharis pilularis* ssp. *consanguinea* invaded annual grassland in California, this has allowed herbivorous mammals to colonize by providing cover from predators. This in turn led to the removal of herbaceous cover and facilitated the transition from grassland to shrubland (Hobbs & Mooney, 1986).

Environment-mediated plant-animal interactions can thus result from the activities of either plants or animals. Up until now, plant-animal interactions have been regarded as predominantly population or community phenomena. The inclusion of environment-mediated interactions illustrates that plant-animal relationships are also important at the ecosystem level and emphasizes their role in ecosystem functioning.

References

Austin, M.P., Groves, R.H., Fresco, L.M.F. & Kaye, P.E. 1985. Relative growth of six thistle species along a nutrient gradient with multispecies competition. J. Ecol. 73: 667–684.

Bandoli, J.H. 1981. Factors influence seasonal burrowing activity in the pocket gopher. J. Mammal. 62: 293–303.

Beattie, A. 1989. The effects of ants on grasslands. In: Huenneke, L.F. & Mooney, H.A. (eds), Grassland Structure and Function, Kluwer, Dordrecht, The Netherlands, pp. 105–116.

Berg, R.Y. 1966. Seed dispersal of *Dendromecon*: its ecologic, evolutionary, and taxonomic significance. Amer. J. Bot. 53: 61–73.

Buckley, R.C. 1982. Ant-plant interactions in Australia, Dr W. Junk, The Hague.

Chiariello, N., Hickman, J.C. & Mooney, H.A. 1982. Endomycorrhizal role for interspecific transfer of phosphorus in a community of annual plants. Science 217: 941–943.

Contreras, L.C. & Gutiérrez, J. 1991. Effects of the subterranean herbivorous rodent *Spalacopus cyanus* on herbaceous vegetation in arid coastal Chile. Oecologia 87: 106–109.

Cook, J.B. 1939. Pocket gophers spread Canada thistle. Calif. Dept. Agriculture Bull. 28: 142–143.

Cox, G.W., Contreras, L.C. & Milewski, A.V. Role of fossorial animals in community structure and energetics of mediterranean ecosystems. In: Arroyo, M.T.K., Zedler, P.H. & Fox, M.D. (eds), Ecology of Convergent Ecosystems: Mediterranean-Climate Ecosystems of Chile, California and Australia, Springer Verlag, New York. In press.

Davies, K.C. & Jarvis, J.V.M. 1986. The burrow systems and burrowing dynamics of the mole-rats *Bathyergus svillus* and *Cryptomys hottentotus* in the fynbos of the south-western Cape, South Africa. J. Zool. 209: 125–147.

Galil, J. 1967. On the dispersal of the bulbs of *Oxalis cernua* Thurnb. by mole-rats (*Spalax ehrenbergi* Nehring). J. Ecol. 55: 787–792.

Goeden, R.D. 1971. The phytophagous insect fauna of milk thistle in Southern California. J. Econ. Entomol. 646: 1101–1104.

Hobbs, R.J. 1985. Harvester and foraging ants and plant species distribution in annual grassland. Oecologia (Berlin) 67: 519–523.

Hobbs, R.J., Gulmon, S.L., Hobbs, V.J. & Mooney, H.A. 1988, Effects of fertiliser addition and subsequent gopher disturbance on a serpentine annual grassland community. Oecologia, 75: 291–295.

Hobbs, R.J. & Mooney, H.A. 1985. Community and population dynamics of serpentine grassland annuals in relation to gopher disturbance. Oecologia (Berlin) 67: 342–351.

Hobbs, R.J. & Mooney, H.A. 1986. Community changes following shrub invasion of grassland. Oecologia (Berlin) 70: 508–513.

Hobbs, R.J. & Mooney, H.A. 1991. Effects of rainfall variability and gopher disturbance on serpentine annual grassland dynamics. Ecology 72: 59–68.

Huntly, N. 1991. Herbivores and the dynamics of communities and ecosytems. Annu. Rev. of Ecol. and Syst. 22: 477–503.

Huntly, N. & Inouye, R. 1988. Pocket gophers in ecosystems: patterns and mechanisms. BioScience 38: 786–793.

Koide, R.T. & Mooney, H.A. 1987. Spatial variation in inoculum potential of vesicular-arbuscular mycorrhizal fungi caused by formation of gopher mounds. New Phytol. 107: 173–182.

Koide, R.T., Huenneke, L.F. & Mooney, H.A. 1987. Gopher mound soil reduces growth and affects ion uptake of two annual grassland species. Oecologia (Berlin) 72: 284–290.

Landsberg, J., Morse, J. & Khanna, P. 1990. Tree dieback and insect dynamics in remnants of native woodlands on farms. Proc. Ecol. Soc. Aust. 16:149–165.

Lovegrove, B.P. & Jarvis, J.V.M. 1986. Coevolution between mole-rats (Bathyergidae) and a geophyte *Micranthus* (Tridaceae). Cimbebasia 8: 79–85.

Matthei, O. 1963. Manual Ilustrado de las Malezas de la Provincia de Ñuble. Concepcion: Universidad de Concepcion, Escuela de Agronomia.

McNab, B.K. 1966. The metabolism of fossorial rodents: a study of convergence. Ecology 47: 712–733.

Michael, P.W. 1968. Thistles in south-eastern Australia – some ecological and economic considerations. In: First Victorian Weeds Conference, Melbourne, pp. 4.12–4.16.

Mielke, H.W. 1977. Mound building by pocket gophers (*Geomyidae*): their impact on soils and vegetation in North America. J. Biogeogr. 4: 171–180.

Milewski, A.V. & Bond, W.J. 1982. Convergence of myrmecochory in mediterranean Australia and South Africa. In: Buckley, R.C. (ed.), Ant-plant Interactions in Australia, Dr. W. Junk, The Hague, pp. 89–98.

Miller, M.A. 1948. Seasonal trends in burrowing of pocket gophers (*Thomomys*). J. Mammal. 29: 38–44.

Naiman, R.J. 1988. Animal influences on ecosystem dynamics. BioSci. 38: 750–752.

Nevo, E. 1979. Adaptive convergence and divergence of subterranean mammals. Ann. Rev. Ecol. Syst. 10: 269–308.
Pimm, S.L. 1982. Food webs. Chapman & Hall, London.
Proctor, J. & Whitten, K. 1971. A population of the valley pocket gopher (*Thomomys bottae*) on a serpentine soil. Am. Midl. Nat. 78: 176–179.
Reig, O.A. 1970. Ecological notes on the fossorial octodont rodent *Spalacopus cyanus* (Molina). J. Mammal. 51: 592–601.
Shortridge, G.C. 1934. The Mammals of South West Africa, William Heinemann Ltd., London.
Singer, M.C. & Ehrlich, P.R. 1979. Population dynamics of the checkerspot butterfly *Euphydryas editha*. Fortscher. Zool. 25: 53–60.

CHAPTER 8

Triangular trophic relationships in Mediterranean-climate Western Australia

BYRON B. LAMONT
School of Environmental Biology, Curtin University, PO Box U 1987, Perth, WA 6001, Australia

Key words: biological control, carnivorous plants, herbivory, mutualism, mycorrhizas, trophic relations

Abstract. The important role of indirect relations between organisms is highlighted. Carnaby's black cockatoo acts as an agent of biological control by selectively seeking moth larvae from flower heads of the rare plant species *Banksia tricuspis*. A burrowing marsupial consumes and disperses the spore bodies of mycorrhizal fungi essential for the growth of major plant species in eucalypt forest. The digestive fluid of carnivorous plants may support a wide range of microorganisms and small animals. Some wingless flies and bugs compete with the host plant for captured prey.

Nomenclature: follows Greene (1985). Nomenclature for animals is as given in the original papers.

Introduction

The food-gathering behaviours of individual species (e.g. carnivory, herbivory) establish definite relationships between species (e.g. predator-prey, mutualistic). Often the interaction between a pair of species is modified by others, called indirect effects (Strauss, 1991). Three is clearly the minimum number of species required for indirect effects to operate, and is explored here in the context of southwestern Australia. Two examples deal with trophic interactions (producer-herbivore-carnivore, producer-symbiont-herbivore). The third examines competition within a trophic level that affects a second trophic level (prey-carnivorous insect *vs* carnivorous plant). These examples demonstrate the delicate and subtle relationships that exist in natural ecosystems and show the futility of concentrating on conservation of individual species at the expense of the ecosystem.

Plant-insect-bird triangle

Banksia tricuspis (Proteaceae) is a gazetted rare species in Western Australia which is restricted to a single area of 50 km^2 in the Lesueur National Park. The aim of this study was to understand the biology of this species as a basis for management practices suitable for conserving it. Most work has concentrated therefore on the reproductive biology and fire ecology of *B. tricuspis* (Lamont & Van Leeuwen, 1988). Early in our study we found 77% of the axes of all flower heads were destroyed by parrots and/or insect grubs from a sample of 50 trees. None of the damaged heads set fruit. The main bird predator was Carnaby's black cockatoo, *Calyptorhynchus funereus latirostris*. This cockatoo, which mainly feeds on seeds and insects (Saunders, 1980), was seeking and consuming the axis-boring grubs, larvae of the moth *Arthrophora* sp. (Tortricidae). It is possible that the birds remove some larvae before they enter the axes, as 40% of heads caged to prevent access to birds became invaded by moth larvae, while only 20% of controls did. Most (89%) of the flower heads bearing grubs were destroyed by birds, while 58% lacking grubs were not attacked and set fruit later.

The following year the exercise was repeated on 300 trees (Fig.8.1). This time, 71% were dam-

M. Arianoutsou and R.H. Groves, Plant-Animal Interactions in Mediterranean-Type Ecosystems, 83–89, 1994.

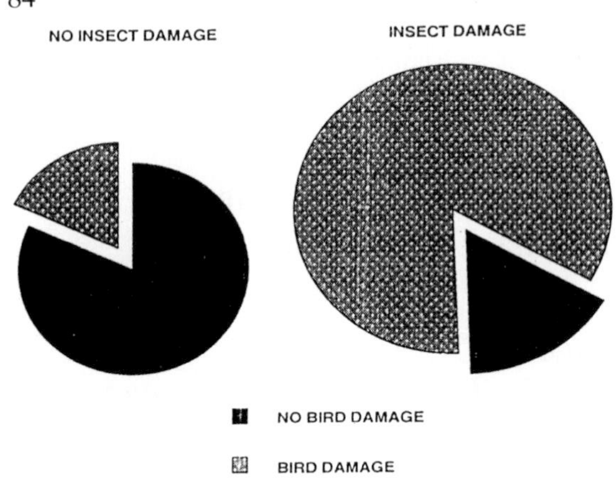

Fig. 8.1. Destruction of the rachis of flower heads of *Banksia tricuspis* by moth larvae and cockatoos. Contingency analysis: $\chi^2 = 1921$, $p \ll 0.001$ (from S. Van Leeuwen, pers. comm.).

aged. The proportion of these that was bird-damaged only was 9%, while the rest were insect-damaged. Of the insect-damaged heads, 84% were also bird-damaged and the tunnels lacked larvae. Just as the moth selectively lays its eggs on trees with most heads, so cockatoos visited those trees preferentially (Lamont & Van Leeuwen, 1988). In consuming 84% of the insects the birds caused an increase in heads destroyed of only 6%.

As a higher order consumer, the cockatoo acts as an agent of biological control, presumably increasing the fitness of the *Banksia* by enabling it to set extra seeds in subsequent crops through reduced abundance of the grubs. The food web (Fig. 8.2) shows a form of mutualism with the reward from the *Banksia* dependent on the presence of the moth, and in whose absence there is no relationship – an example of facultative mutualism. The relationship is not unique as both the moth and cockatoo behave in a similar way on other *Banksia* species in the region (Saunders, 1980; Scott, 1982).

Plant-fungus-mammal triangle

A small marsupial, the woylie, *Bettongia penicillata*, is the major consumer of the underground sporocarps of at least 18 species of higher fungi in eucalypt forest in Western Australia (Lamont *et al.*, 1985). Consumption increases after fire when little other food is available. Most spores in the faecal pellets belong to the mycorrhizal genus *Mesophellia* (Malajczuk *et al.*, 1987). When fresh pellets were applied to seedlings of *Gastrolobium bilobum* (Fabaceae) and *Eucalyptus calophylla* (Myrtaceae) in autoclaval soil they formed more ectomycorrhizal rootlets than the controls in non-autoclaved soil (Table 8.1). Seedlings in autoclaved soil receiving autoclaved pellets formed no mycorrhizas and plant mass reached only 70% of the eucalypt seedlings treated with fresh pellets and 11% for the pea seedlings. Furthemore, application of fresh spores of two *Mesophellia* species to the seedlings produced neither mycorrhizas nor growth responses. The most likely explanation is that digestion by the woylie facilitates germination of the spores, although this does not always appear necessary (Malajczuk *et al.*, 1987). Our more recent work on *Gastrolobium* has shown that the pellets are likely to in-

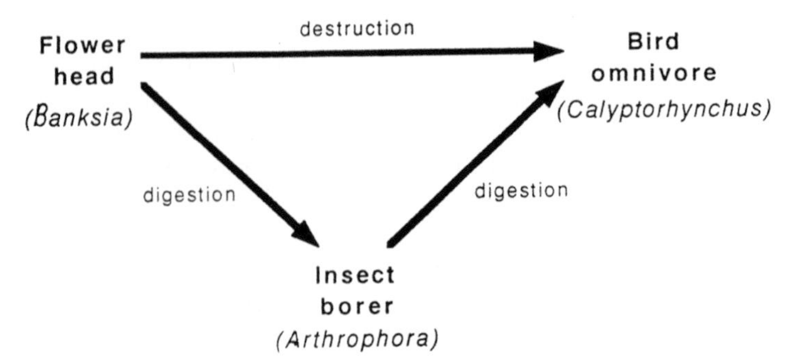

Fig. 8.2. Food web between *Banksia tricuspis* (producer), *Arthrophora* sp. (herbivore) and *Calyptorhynchus funereus* (carnivore) with the flow of matter indicated by the direction of the arrows.

Table 8.1. Effect of woylie faeces and *Mesophellia* spores on mycorrhizal status and growth of *Eucalyptus calophylla* and *Gastrolobium bilobum* after 6 months. Results with different letters significantly different at $p < 0.01$. (adapted from Lamont *et al.*, 1985)

Treatment	Mycorrhizal roots (%)		Total growth (mg)	
	Eucalypt	Pea	Eucalypt	Pea
Untreated soil (control)	24^b	28^b	575^c	859^a
Autoclaved soil + fresh faeces	56^a	89^a	1359^a	759^a
Autoclaved soil + autoclaved faeces	0^c	0^c	951^b	81^b
Autoclaved soil + fungal spores	0^c	0^c	736^c	56^b

crease the numbers of mycorrhizal types in a range of soils and fire histories from three to five (L. Stewart and B. Lamont, unpub.).

The woody plants are the producers whose growth, and probably survival in the case of the pea, depends on a physiological relationship (symbiosis, *sensu* Cushman & Beattie, 1991) with a fungal heterotroph. Restoration of these fungi after fire is vital for re-establishment of species such as the *Gastrolobium*, as it is killed by fire. Not only does consumption of sporocarps increase after fire, but the woylie may travel up to 3 km overnight and moves from one burnt patch to another (Lamont *et al.*, 1985). This movement gives the woylie the capacity to be an effective dispersal and restoration agent, but the extent to which the soil is sterilized by fire remains uncertain: it probably depends on its intensity (Malajczuk & Hingston, 1981). In addition, this marsupial appears to predispose the spores to germinate – the consumer thus has a mutualistic association with its food source. The woylie has both direct and indirect effects on the population dynamics of mycorrhizal species. As a ‘fire weed’ the *Gastrolobium* and other mycorrhizal peas form dense thickets after fires of high intensity: these thickets serve as essential shelters from carnivores and nesting sites for the woylie (Christensen & Leftwich, 1980). The woylie has the strange habit of burying *Gastrolobium* seeds in ‘caches’ (Christensen, 1980). To what extent woylies eat the germinants is unknown, although they have exceptionally high tolerance of the toxin fluoroacetate (10–80) present in *Gastrolobium* and other local poison peas. While the woylie is currently the major consumer, other mammals are also known to consume fungi – thus there are a number of alternative partners at each

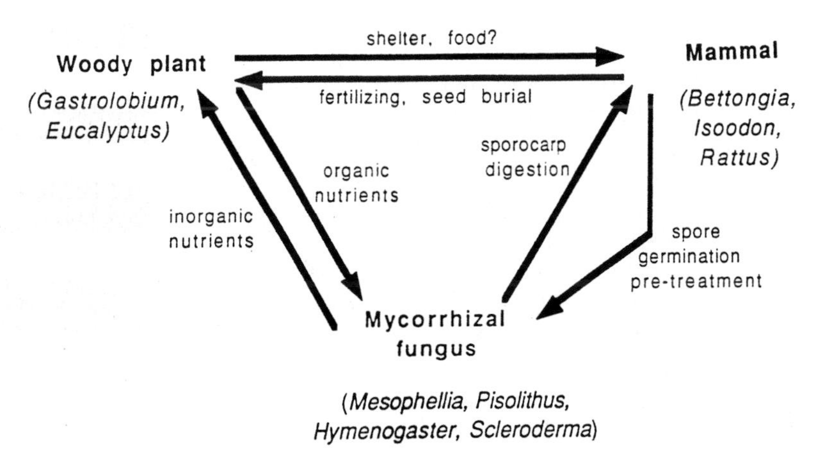

Fig. 8.3. Food web between woody plants (producers), mycorrhizal fungi (heterotrophs), and *Bettongia penicillata* and other mycophagous mammals – all relationships are mutualistic.

Table 8.2. Animals present in the pitcher fluid of *Cephalotus follicularis* which occurs along the south coast of Western Australia (collated from Hamilton, 1904; Clarke, 1985; Yeates, 1990)

Common name	Broad group	Specific group	Trophic function
Ant	Hymenoptera	*Iridomyrmex conifer*	Prey – digested
Ant	Hymenoptera	other species	Prey – digested
Beetle	Coleoptera	? (adult)	Prey – digested
Moth	Lepidoptera	? (adult)	Prey – digested
Mosquito	Diptera	? (adult)	Prey – digested
Copepod	Copepoda	*Elaphoidella* sp.	Detritus feeder
Eelworm	Nematoda	?	Generalist feeder
Mite	Acarina, Oribatidae	?	Generalist feeder?
Earthworm	Oligochaeta	?	Detritus feeder
Fly	Diptera, Micropezidae	*Badisis ambulans* (larva)	Carrion feeder
Fly	Diptera, Ceratopogonidae	*Dasyhelea* sp. (larva)	Carrion feeder

trophic level in the partnership. All relationships in the triangle are reciprocal – from indirect to obligate mutualism (Fig. 8.3).

Plant-arthropod-insect triangle

Carnivorous plants are better represented in Mediterranean climate areas of Western Australia than anywhere else in the world (Lamont, 1982). Many arthropods, mainly small insects, are captured by glandular leaves produced by these herbs. They are digested by enzymes and organic acids secreted by the glands and their associated bacteria to supplement the plant's mineral nutrient supply (reviewed in Lamont, 1982). Some animals use the digestive fluid in the Albany pitcher plant, *Cephalotus follicularis* (Cephalotaceae), opportunistically as a feeding ground (Table 8.2). Most of these are detritus feeders taking advantage of the partly decomposed remains of prey, algae, fungi, bacteria, protists and microscopic animals present in the fluid (Fig. 8.4, Clarke, 1985). Of particular interest are two carrion-feeding flies – a small proportion of the pitchers contain their eggs, larvae or pupae at any time of the year. Yeates (1990) considers that *Badisis ambulans* breeds only in the pitchers. This wingless fly appears to mimic the ant, *Iridomyrmex conifer*, the most common prey of the pitcher plant. Ants feed from nectaries distributed over much of the surface of the pitcher, but especially on the slippery teeth over the rim (Parkes & Hallam, 1984). Because of this reward, and the fact that only a tiny fraction of visitors slips in,

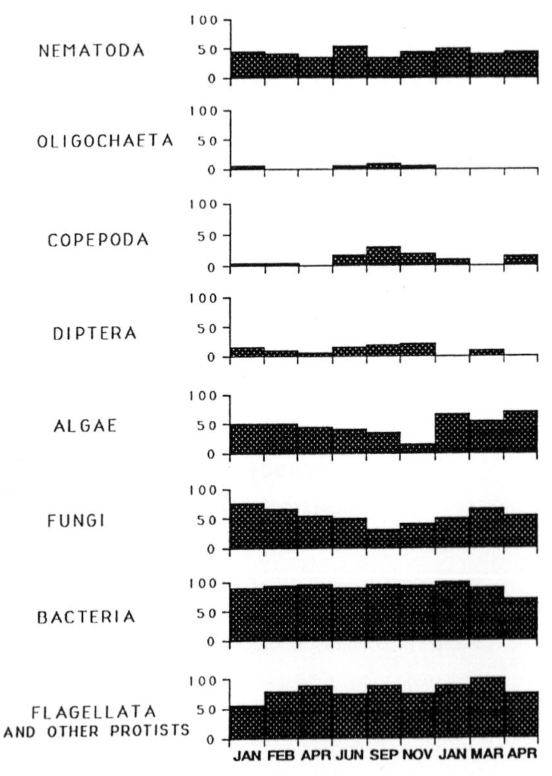

Fig. 8.4. Frequency of occurrence of major groups of organisms in pitchers of *Cephalotus follicularis* over 15 months. Decomposers are the most abundant groups with worms, copepods and insect larvae sometimes present. (collated from Clarke, 1985).

Fig. 8.5. Two nymphs of *Setocaris bybliphilus* on glandular stem of *Byblis gigantea*. These nymphs feed on prey captured by the tentacles of the plant but much remains to be learnt about their relationships with the host plant. Nymphs are about 4 mm long. Photo: B. Lamont.

Joel (1988) concludes that pitcher plants are neither mimics nor commensalists – they provide in fact a good example of mutualism. If the fly lures even more ants into the pitcher its relationship with the plant could also be mutualistic, rather than commensal as normally held, but it is too fitful to be obligate.

Some insects, such as the scorpion fly, *Harpabittacus australis* (Russell, 1953), risk capture by the tentacles of sundews (*Drosera*, *Byblis*) by indulging in herbivory, mucilage sipping or feeding on the captured prey. Some bugs have evolved an obligate association with sundews, however (China, 1953; Taylor, 1989). So far, three taxa have been recognized (*Cyrtopeltis droserae*, *C. russellii*, *Setocoris bybliphilus*) in the essentially sap-sucking family Miridae, sub-family Phylinae, tribe Dicyphini (Fig. 8.5). All are probably more

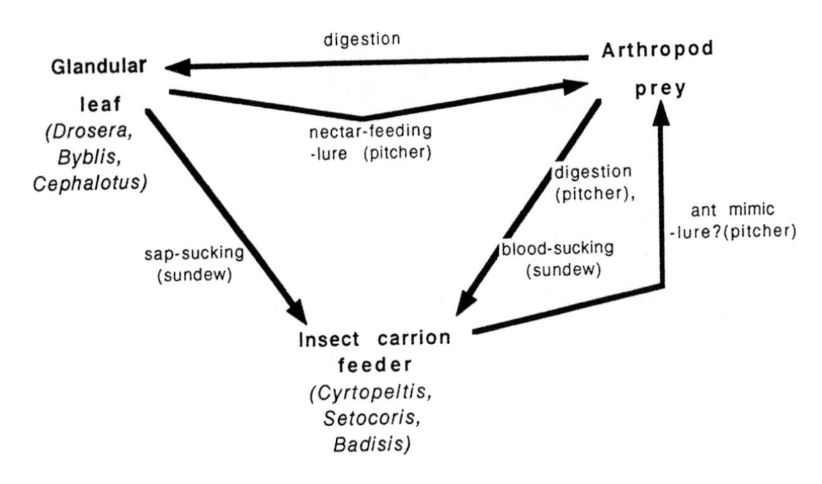

Fig. 8.6. Food web for carnivorous species in Mediterranean Western Australia showing their relationship to arthropod prey and carrion-feeding opportunist insects. Traditionally regarded as commensal, mutualistic relations are now becoming evident.

correctly placed in *Setocoris* and many more species exist (G. Cassis, pers. comm.). They retain their sap-sucking ability (Russell, 1953; G. Cassis, pers. comm.) but more importantly suck the blood of prey captured by the sundews. Like the animals living in the pitcher fluid, they compete with the host plant for the same resource. Since this is likely to have negligible effect on the fitness of the sundew, the relationship is currently described as commensal. As with the ant-mimicking fly, mutualism remains possible, however – perhaps the bugs also prey on larval herbivores as do other hemipterans? The food web for both groups of carnivorous plants is described in Fig. 8.6.

Discussion

I have outlined a few examples of three-way relationships between plants and animals in a Mediterranean climate region of Australia. If any one component were absent, the relationship between the other two would be quite different. For conservation purposes it is essential to know the functional relationships of the species of interest to other species. Some are facultative, as with the black cockatoo feeding on the moth larvae in the flower heads of *Banksia tricuspis*: in the absence of the moth, the bird would probably not visit the flowers. Others are obligate, as with the mycorrhizal fungi associated with the legume *Gastrolobium bilobum*: without the microsymbiont, seedlings of this species would not survive. Whether the marsupial dispersers of the spores are also obligate depends on a) whether spores of all fungi must pass through their body before they will germinate – this seems unlikely; b) if this is not true, whether the spores can be distributed passively – this would appear inefficient at best; and c) to what extent the soil is sterilized (mycorrhizal fungi are killed) by fires – this will depend on the intensity and patchiness of the fire.

Another aspect is the extent to which substitutes (surrogates) exist at each trophic level. For example, to what extent can the assassin bugs switch from one host *Drosera* species to another? – there are almost 50 species of these plants in southwestern Australia. Since the bugs are using the plant merely as a source of prey there should be no impediment to their switching to a new host species in the vicinity. The timing would be important – adults could fly to new hosts in the community, while nymphs are confined to the plants on or near where they hatch.

The quantitative aspects of the relationship may be as important as its qualitative nature (Lamont, 1992). This is well-illustrated with mycorrhizal fungi, where there may be a difference of an order of magnitude in effects of different species on enhancing growth of the host species (Bougher *et al.*, 1990). This throws a new light on the search for surrogates – if some participate in the pathway but at such a slow or fitfull rate as to threaten the viability of the associated species, directly or indirectly, then it is not a substitute at all. This is probably true for the corella as a replacement for the black cockatoo in controlling moth larvae, and other small mammals as a substitute for the woylie as fungal dispersal agents.

The 'keystone' role of a given species is difficult to examine experimentally – it involves its removal from the ecosystem while the ecological effects of such removal may not be evident for hundreds of years. Revelation of these intricate relationships is just the 'tip of the iceberg' – the more species examined, the geometrically more complex the relationships revealed. In the face of escalating threats to biodiversity in all regions of Mediterranean climate, the elucidation of much intricate relationships presents an urgent challenge to ecologists.

Acknowledgements

Sue Radford processed the figures for publication. Bert Main gave me permission to use material from Sally Clarke's thesis.

References

Bougher, N.L., Grove, T.S. & Malajczuk, N. 1990. Growth and phosphorus acquisition of karri (*Eucalyptus diversicolor* F. Muell.) seedlings inoculated with ectomycorrhizal fungi in relation to phosphorus supply. New Phytol. 114: 77–85.

China, W.E. 1953. Two new species of the genus *Cytopeltis* (Hemiptera) associated with sundews In Western Australia. West. Aust. Nat. 4: 1–8.

Christensen, P.E. 1980. The biology of *Bettongia penicillata* Gray, 1837, and *Macrcpus eugenii* (Desmarest, 1817) in relation to fire. Forests Dept. West. Aust. Bull. No. 91.

Christensen, P. & Leftwich, T. 1980. Observations on the nest-building habits of the brush-tailed rat-kangaroo or woylie (*Bettongia penicillata*). J. Roy. Soc. West. Aust. 63: 33–38.

Clarke, S.A. 1985. Demographic aspects of the pitcher of *Cephalotus follicularis* (Labill.): and development of the contained community. Ph.D. Thesis, Univ. West. Aust.

Cushman, J.H. & Beattie, A.J. 1991. Mutualisms: assessing the benefits to hosts and visitors. Trends Ecol. Evol. 6: 193–195.

Hamilton, A.G. 1904. Notes on the West Australian pitcher plant (*Cephalotus follicularis* Labill.). Proc. Linn. Soc. N.S.W. 29: 36–56.

Green, J.W. 1985. Census of the vascular plants of Western Australia. West. Aust. Dept. Agric., South Perth.

Joel, D. 1988. Mimicry and mutualism in carnivorous pitcher plants (Sarraceniaceae, Nepenthaceae, Cephalotaceae, Bromeliaceae). Biol. J. Linn. Soc. 35: 185–197.

Lamont, B. 1982. Mechanisms for enhancing nutrient uptake in plants, with particular reference to mediterranean South Africa and Western Australia. Bot. Rev. 48: 597–689.

Lamont, B.B. 1992. Functional interactions within plants – the contribution of keystone and other species to biodiversity. In: Hobbs, R.J. (ed). Biodiversity of Mediterranean Ecosystems in Australia, Surrey Beatty & Sons, Chipping Norton, NSW, pp. 95–127.

Lamont, B.B. & Van Leeuwen, S.J. 1988. Seed production and mortality in a rare *Banksia* species. J. Appl. Ecol. 25: 551–559.

Lamont, B.B., Ralph, C.S. & Christensen, P.E. 1985. Mycophagous marsupials as dispersal agents for ectomycorrhizal fungi on *Eucalyptus calophylla* and *Gastrolobium bilobum*. New Phytol. 101: 651–656.

Malajczuk, N. & Hingston, F.J. 1981. Ectomycorrhizae associated with jarrah. Aust. J. Bot. 29: 453–462.

Malajczuk, N., Trappe, J.M. & Molina, R. 1987. Interrelationships among some ectomycorrhizal trees, hypogeous fungi and small mammals: Western Australian and northwestern American parallels. Aust. J. Ecol. 12: 53–55.

Parkes, D.M. & Hallam, N.D. 1984. Adaptation for carnivory in the West Australian pitcher plant *Cephalotus follicularis* Labill. Aust. J. Bot. 32: 595–604.

Russell, M.C. 1953. Notes on insects associated with sundews (*Drosera*) at Lesmurdie. West. Aust. Nat. 4: 9–12.

Saunders, D.A. 1980. Food and movements of the short-billed form of the white-tailed black cockatoo. Aust. Wild. Res. 7: 257–269.

Scott, J.K. 1982. The impact of destructive insects on reproduction in six species of *Banksia* L.f. (Proteaceae). Aust. J. Zool. 30: 901–921.

Strauss, S.Y. 1991. Indirect effects in community ecology: their definition, study and importance. Trends Ecol. Evol. 6: 206–210.

Taylor, J. 1989. Flower Power in the Australian Bush and Garden. Kangaroo Press, Kenthurst. NSW.

Yeates, D. 1990. When the pitcher pays. Aust. Nat. Hist. 23: 512–513.

PART FOUR

Herbivory

CHAPTER 9

Has intensive grazing by domestic livestock degraded Mediterranean Basin rangelands?

NO'AM G. SELIGMAN and AVI PEREVOLOTSKY
Department of Agronomy and Natural Resources, Agricultural Research Organization, Volcani Center, P.O. Box 6, Bet Dagan 50-250, Israel

Key words: grazing, Mediterranean Basin rangelands, vegetation

Abstract. The response to grazing of the dominant vegetation types on rangelands in the Mediterranean Basin is reviewed. These vegetation communities are not only well adapted to heavy grazing, but low grazing pressure can have undesirable ecological and management consequences. It is suggested that a new approach to management of Mediterranean Basin rangelands should take into account the special characteristics of the vegetation and the rangeland habitats, the specific human and biological history of the region, and the changing socio-economic environment.

Introduction

The rough and rocky rangelands that surround the Mediterranean Basin are viewed usually as a degraded landscape. For instance, Foran *et al.* (1989) stated that "The rangelands of the Middle East have been used and abused for more than 5000 years. In many areas there is little left". This view was also held by Plato, who, 2,400 years BP, described the hills around Athens to be '. . . like the skeleton of an old man, all the fat and soft earth wasted away and only the bare framework of the land being left' (cited by Attenborough, 1987). Indeed, the so-called 'cradle of civilization' has borne the brunt of intensive human activity for millennia. Wood-cutting, clearing of land for cultivation and settlement, fire, and, in recent years, 'neotechnical factors' (pollution, biocides), have all shaped the ever-changing landscape (Naveh & Leiberman, 1984; Attenborough, 1987). Additional to all these factors, however, overgrazing has been singled out as being especially pernicious (Naveh, 1975; Tomaselli, 1977; Le Houerou, 1981; Thirgood, 1981). Goats, in particular, have been widely incriminated: '. . . The goats consume every seedling that sprouts and every leaf that unfurls so the land remains barren.' (Attenborough, 1987). Reid (1908) wrote in a report about Cyprus (as cited by Thirgood (1987): 'This mischief, erosion and water flooding, is due mainly to the enormous herds of goats which destroy the young trees. The goat is a greater curse . . . than the locust. The locust destroyed the vegetation for a single season, the goat destroys the vegetation permanently'. Environmentalist lobbies have been pressing for the reduction or elimination of goat grazing (Papanastasis, 1986). In Israel, for example, a law was passed in 1960 that severely restricted goat husbandry on rangelands. In Cyprus such a law was decreed in 1913 but with little effect (Thirgood, 1987). A MAB poster cites only 'overgrazing' as an example of misuse in marginal lands that 'can set off a chain of events leading to soil degradation' (Naveh & Lieberman, 1984). Mrs. Niki Goulandris, co-founder of the Goulandris Museum of Natural History in Kifissia, Greece, recipient of the 1991 'Woman of Europe' award and champion of environmental causes, has stated that '. . . apart from destruction by man, the worst threat [to the Mediterranean forests] comes from goats' (Sakellariou, 1991).

With such wide and august consensus, it may seem audacious to suggest that grazing by dom-

M. Arianoutsou and R.H. Groves, Plant-Animal Interactions in Mediterranean-Type Ecosystems, 93–103, 1994.

estic ruminants is seldom irreversibly destructive to landscape values. Yet, there is inadequate evidence to substantiate the claim that grazing by domestic ruminants is a major cause of rangeland degradation or of desertification, even in stressed traditional pastoral regions (Sandford, 1986; Ellis & Swift, 1988; Perevolotsky, 1992). We will argue in this chapter that the popular consensus is too simplistic and does injustice to a far more complex situation. It is important to clarify the issue because with major change in land use taking place all over the Mediterranean Basin, it is increasingly appropriate to review questions and problems of ecosystem management in the region. Is the 'guilt for degradation', real or apparent, fairly apportioned? Has this allocation any management or other implications within the Mediterranean Basin and in other Mediterranean-climate ecosystems.

The rangeland vegetation of the Mediterranean Basin

The terrestrial vegetation of the Mediterranean Basin has been a source of human and animal food, building material and fuel for millennia and the evolution of the vegetation under a regime of grazing and human exploitation has been well documented (Di Castri & Mooney, 1973). There have been periods of major change in the landscape as primaeval forest was cleared for cultivation, habitation or industrial purposes. 'When states went to war, entire forests were devastated to provide the armies with vehicles and the navies with ships' (Attenborough, 1987). Periods of development were followed by neglect when civilizations declined, terraces were abandoned and soil erosion denuded the previously productive hillsides (Naveh & Dan, 1973). Land unsuitable for cultivation has been traditionally used as range. This negative definition depends on what is deemed suitable for cultivation, a classification that has changed with the varying fortunes of civilization in the region. Consequently, much land that today is rangeland, was cultivated in the past. But throughout the ages, the rangelands in the Mediterranean Basin have been rocky, sloping lands with patchy soils and varying cover of herbaceous and woody vegetation. The view of these lands as degraded forests gave rise, in Israel, to an aggressive conservation movement that, together with a drastic reduction in numbers of grazing livestock after 1948, led to a dramatic recovery of the sclerophyllous woodland. In Upper Galilee, dense thickets of oak (*Quercus calliprinos*) now stand in sharp contrast to the bare, rocky landscape across the Lebanese border that still maintains large numbers of productive livestock. The woody vegetation cover has become lush, but in many places the stand is so thick that access is now severely restricted, wildlife is scarce and fire hazards have increased sharply. Today, 40 years later and with a more informed acquaintance with the situation, the clear distinction between the benefits of conservation and the negative effects of heavy grazing is becoming blurred (Perevolotsky, 1991b). Even though '. . . our present knowledge on the long-term effects of different modes of grazing . . . is only fragmentary . . .' (Naveh, 1971), there is a substantial amount of information that can enable us to define criteria and judge more objectively the multiple dimensions of land use in the Mediterranean region.

Criteria for ecosystem degradation

The plant-animal interactions in ecology are biological events with no value judgements involved. 'Degradation' is a change of state with an implied negative value assessment. Also, grazing of domestic animals on public domain is not only a biological process but also a socio-political issue that involves the relationship between different sectors of the community and the landscape. Consequently, whilst a herder will regard rangeland vegetation as forage for his livestock – and an oak grazed down to a dense dwarf shrub is simply a well exploited forage plant – a forester or environmentalist will regard the domestic ruminant as a pest that threatens the forest or restricts the proper development of the woody vegetation. No doubt, the effects of herbivores on vegetation can be dramatic, but, as in the following classic example cited by Crawley (1983), the effect on species survival is often more apparent than real, viz. 'On an acid heath in Surrey where only small isolated clumps of mature Scots pine trees could be seen, a fence was erected to exclude cattle.

Almost immediately pine saplings sprang up everywhere in abundance. Intrigued by this, Darwin (1859) looked more closely at the 'treeless' heath and 'in one square yard, at a point some hundreds yards distant from one of the old clumps, I counted thirty-two little trees; and one of them, judging from the rings of growth, had during twenty six years tried to raise its head above the stems of the heath, and had failed. No wonder that, as soon as the land was enclosed, it became thickly clothed with vigorously growing young firs'.' Similar situations are common in Mediterranean woodland communities.

Criteria of degradation, therefore, depend on the objectives of the viewer or interested party. Commonly, there is more than one legitimate stakeholder interested in the use of the rangeland and different objectives must often be accommodated on the same area. Whilst multiple-use is a much desired objective (Holechek *et al.*, 1989), it is often difficult to resolve conflicts of interest, especially when there is no clear distinction between the facts and some of the (often-convincing) fictions.

The land use class that is relevant to a review of the effects of intensive grazing by domestic livestock is the area that today, or in the recent past, has been legitimately used by the people of the region as rangeland. This excludes conservation areas (reserves, parks) and some types of commercial forests as well as currently cultivated land. The scale of the grazing effect must also be considered. On many relatively small areas grazing has destroyed the vegetation completely and contributed to habitat degradation, particularly where animals congregate (i.e., at water points, in paddocks for protection at night, on resting sites or along driveways). These are 'sacrifice' areas that normally constitute a relatively small part of the domain that serves as the 'Lebensraum' for the livestock (Stoddard *et al.*, 1975). Some special sites, like riparian areas or springs, may draw heavy livestock concentrations that can cause extensive environmental damage (Holechek *et al.*, 1989). The relevant scale for the present review is the main body of the grazing domain, generally the landscape.

Criteria for judging whether herbivory by domestic livestock has been excessive and has caused ecosystem degradation, should consider both the structure and the function of the ecosystem. Structural aspects include changes in species and habitat diversity, excessive soil loss, landscape and amenity values. Functional aspects include changes in primary and secondary productivity, plasticity, resilience, and proneness to fire.

Structural criteria

Habitat diversity. The open, heavily grazed Mediterranean landscape can appear as uniformly bleak, especially during the dry summer. Closer inspection reveals, however, that Mediterranean plant communities are composed of a 'mosaic of inumerable variants of degradation and regeneration stages' (Walter, 1968). Many elements of the mosaic are on a small scale and are maintained only by control of the woody vegetation, primarily by grazing. Development of a dense shrub or tree overstorey reduces light intensity in the understorey and thereby creates a more uniform habitat where variation in microtopography and soil factors is damped. For instance, the persistence of *Lilium candidum* in the sclerophyllous woodland on Mt. Carmel in Israel is being threatened because it does not thrive in the shade of the encroaching shrubs (A. Dafni, personal communication). Consequently, it is becoming increasingly rare in conservation areas where once it was common when the land was heavily exploited for grazing and woodcutting. There are, of course, different habitats in a conserved woodland compared with an open, heavily grazed landscape, but whether habitat diversity is any greater, is open to question. The biological uniformity of some densely wooded areas that exclude many local species of native fauna and flora (Naveh, 1971) has given rise to the connotation 'green desert' (Sela, 1975).

Species diversity. Where grazing reduces the number of species in a region, it can be defined as excessive ('overgrazing'). Consequently, where species diversity is increased by herbivory, grazing is not excessive, at least in this sense, and 'overgrazing' is not an appropriate term. There are, however, cases where protection from grazing (particularly rabbit grazing) has led to the development of a richer flora and a more continuous cover of vegetation (Watt, 1981). There is

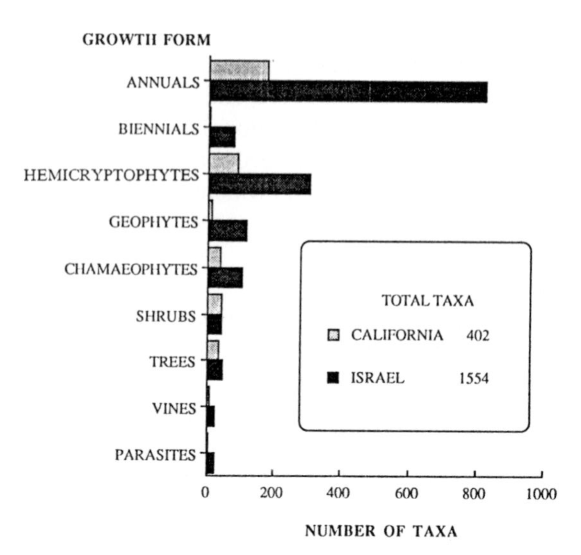

Fig. 9.1. Growth from spectra of chaparral flora in Israel and California (after Shmida, 1981).

ample evidence that moderate grazing increases species diversity (Harper, 1969; Naveh & Whittaker, 1979; Noy-Meir *et al.*, 1989; Noy-Meir & Kaplan, 1990). 'The principal effect of herbivores on plant species richness acts not through animals eating plants to extinction (although this can happen), but through their feeding modifying the competitive abilities of one plant species with another' (Crawley, 1983). Grazing, even heavy grazing, is being recommended to maintain and increase species diversity in Israeli nature reserves (Safriel, 1991).

Species diversity can be evaluated on different scales. Starting with the largest scale, we can compare the diversity of the Mediterranean-type flora in North and South America, where intensive domestic animal grazing is fairly recent, with that in the Mediterranean Basin, where it has a much longer history. Shmida (1981) found that the number of plant species in Israel is nearly four times higher than under similar ecological conditions in California (Fig. 9.1). In addition, more than one-quarter of the Californian species are introduced annuals, mostly from the Mediterranean Basin. At the other end of the scale is a study by Papageorgiou (1979) that compared two islands near Chania, Crete: one 'heavily overgrazed' by the Cretan wild goat, the 'agrimi' (*Capra aegagrus cretica*), and the other, ungrazed. Whilst

Table 9.1. Numbers of species and percent cover for vegetation on a heavily grazed island (Theodoru) and a nearby ungrazed islet (Theodoropoula), Crete (after Papageorgiou, 1979)

Species group	Theodorou		Theodoropoula	
	Number	% cover	Number	% cover
Shrubs	8	88	3	22
Forbs	46	11	10	74
Grasses	4	1	3	4
Total	58	–	16	–

the two islands are not strictly comparable (the grazed one is much larger than the ungrazed one), there are many more species in the 'heavily overgrazed' islet (Table 9.1).

Ungrazed, dense oak thickets in Israel have a lower species diversity than that in open, grazed woodlands (Naveh & Whittaker, 1979). With the development of a dense overstorey of woody species, the diversity of the understorey species is reduced because of both competition with the overstorey species and severe reduction in light intensity (Specht & Specht, 1989; Specht *et al.*, 1990). The reduced understorey diversity cannot be compensated for by the overstorey because the number of overstorey species is invariably much smaller than the number of understorey species (Fig. 9.1). It must be concluded that many Mediterranean rangelands that are 'overgrazed' when judged as forests or natural reserves, are moderately grazed in terms of species diversity.

Rare species often depend on special habitat conditions that can be threatened by excessive grazing. Riparian habitats are particularly sensitive to degradation by heavy grazing (Holechek *et al.*, 1989) and should be suitably protected. On the other hand, there are also rare species, particularly small annuals and geophytes, that seem to persist, and even depend, on a heavily grazed environment.

Water and soil loss. Heavy grazing is commonly assumed to increase runoff and associated soil loss (Naveh & Lieberman, 1984; Holechek *et al.*, 1989). Results of many studies, particularly in the U.S.A., have shown that soil loss on grazed range is greater than on ungrazed range (Stoddard *et al.*, 1975; Holechek *et al.*, 1989). Grazing is therefore restricted on critical watersheds in the U.S.A. Thirgood (1987) declared that 'in the final analysis, particularly in the mountains of Cyprus, erosion has been directly related to goat browsing.'

In the Mediterranean Basin many, if not most, terrestrial rangeland ecosystems are on fractured, karstic, rocky uplands covered with a layer of well-structured soil of varying depth and protected in patches and pockets between the rock outcrops. Consequently, runoff and associated erosion from such watersheds are low, most runoff originating from cultivated lands during occasional heavy storms, particularly when the soil is bare at the beginning of the rainy season. It has been estimated that from cultivated terraces on Mount Carmel in Israel, soil loss is less than 5 mm in 100 years (Seginer *et al.*, 1963). In an experiment on prescribed burning for forest plantation management in Israel, erosion in the burned plots was no greater than in the control plots (Y. Zohar, pers. comm.). Erosion is reduced by cover of vegetation and litter. Grazing, even heavy grazing, seldom leaves the soil as bare as cultivation or burning does. The impact of grazing on soil erosion is, therefore, less drastic than the effects of either burning or cultivation, particularly on the karstic rangeland soils. There are more sensitive soils, however, particularly on schist formations, where excessive grazing can lead to greater runoff and accelerated soil degradation. Such areas are common on the uplands in Greece, Italy, France and Spain.

The predominant landscape forms in the rangelands of the Mediterranean Basin are fashioned from the limestone deposits of the ancient Tethys Sea. On these landforms, the effect on the water balance of grazing the scrub woodlands is not necessarily negative. Heavy woody growth, particularly deep-rooted trees and shrubs that can extract water from the karstic rock-soil complex down to 11 m and below, transpire more than shallow-rooted herbaceous vegetation (Caldwell *et al.*, 1986). In a study on Mount Carmel in Israel, the water use of oak forest in some years was found to be 180 mm more than that of a similar area of herbaceous vegetation (Rozenzweig, 1972). Where recharge of the underground aquifers is important for the regional water balance, a hydrological advantage can be gained by maintaining a low woody cover on Mediterranean rangeland watersheds.

Landscape and amenity value. Forest-clad hillsides can contribute to a balanced environment, ecologically desirable and aesthetically satisfying. This is particularly so where such landscapes are uncommon. In promoting the expansion of forest areas, extreme positions are sometimes adopted, however, with regard to the function of grazing in Mediterranean landscapes. So Mrs. Goulandris (see above) hopes that the 'nibbling of Greece' will come to an end with the EEC ban on subsidies for goat herders, to take effect from 1992, and the gradual dying out of the profession (Sakellariou, 1991). In much of the Mediterranean region, particularly in the oak communities, a dense scrub forest has limited amenity value. Where woodlands are open and accessible, there are far more opportunities for recreation, from hiking and camping to hunting. Where they are closed and dense, they not only have limited amenity value, but are also a fire hazard in a fire-prone environment (Naveh & Lieberman, 1984). Open, rocky, heavily grazed landscapes are commonly described as degraded, but are often attractive recreation areas, with seasons of abundant and diverse flowering.

Whilst large areas in the Mediterranean Basin that are denuded of woody vegetation can be described as degraded forests, many areas are without woody vegetation because of habitat characteristics, particularly edaphic factors that create soil water regimes unfavorable for woody species (Berliner, 1970; Rabinovich-Vin & Orshan, 1974), or because of previous clearing and cultivation (Litav, 1965, 1967). Such areas (the eastern slopes of Galilee, and the Golan Heights) remain grassland even when grazing pressure is reduced.

Functional criteria

Primary production. As a rule, grazing, particularly heavy grazing, reduces primary production in seasonal herbaceous pastures (Noy-Meir, 1978; Holechek *et al.*, 1989). This may not necessarily be the rule in shrubby pastures where the ability to graze is low and allocation of photosynthates to the woody component and to roots is

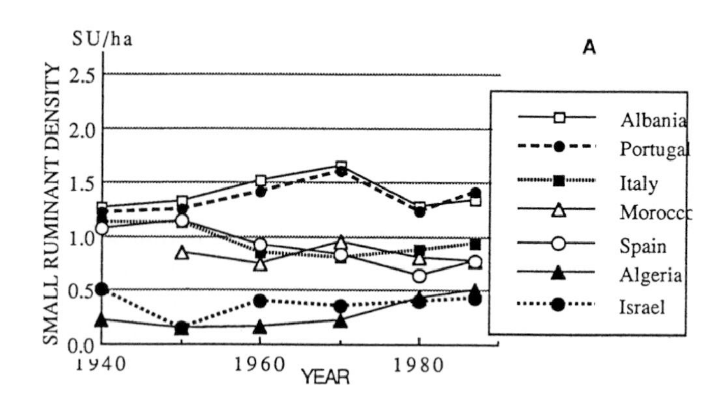
A
SU/ha
SMALL RUMINANT DENSITY
2.5
2.0
1.5
1.0
0.5
0.0
1940
1960
1980
YEAR
Albania
Portugal
Italy
Morocco
Spain
Algeria
Israel

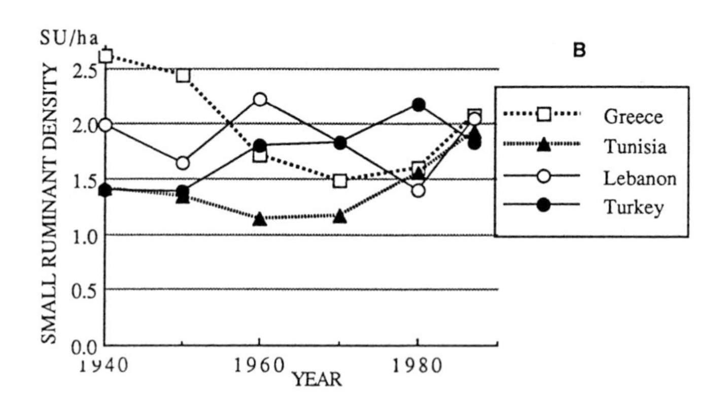
B
SU/ha
SMALL RUMINANT DENSITY
2.5
2.0
1.5
1.0
0.5
0.0
1940
1960
1980
YEAR
Greece
Tunisia
Lebanon
Turkey

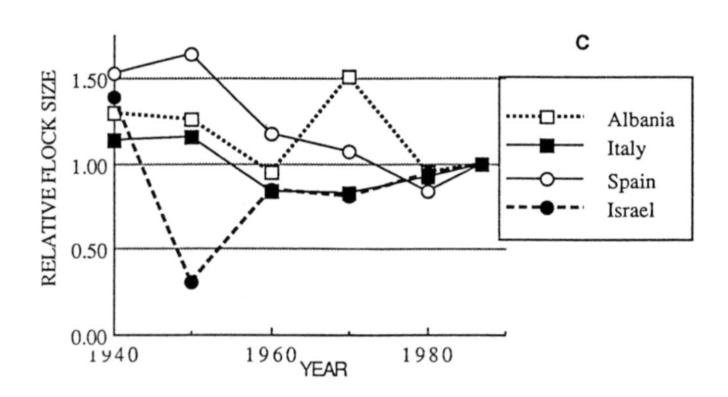
C
RELATIVE FLOCK SIZE
1.50
1.00
0.50
0.00
1940
1960
1980
YEAR
Albania
Italy
Spain
Israel

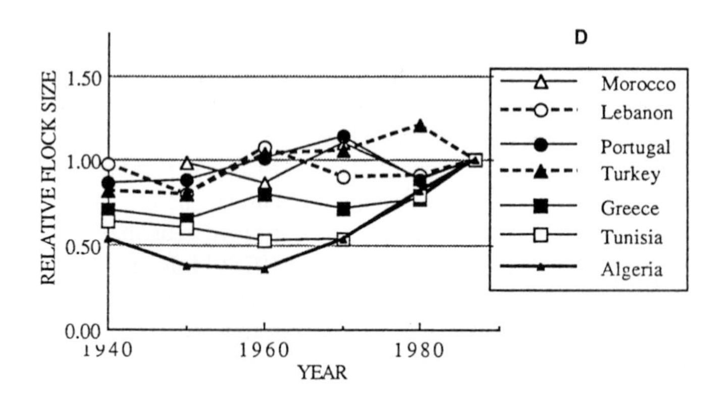
D
RELATIVE FLOCK SIZE
1.50
1.00
0.50
0.00
1940
1960
1980
YEAR
Morocco
Lebanon
Portugal
Turkey
Greece
Tunisia
Algeria

high. Under such conditions, defoliation often produces a vigorous growth response. Intensive clipping of kermes oak increased twig production in Greece (Tsiouvaras *et al.*, 1986; Tsiouvaras, 1987), whilst in the U.S.A., even ten consecutive years of heavy defoliation of *Artemisia tridentata* did not significantly reduce production (Shepherd, 1971). Similar effects have been obtained with other shrub species (Garrison, 1953). The relevant question in the context of this chapter, however, is not whether grazing reduces production in relation to ungrazed vegetation, but is the potential production of the ecosystem reduced by heavy grazing. Results of long-term grazing studies in Israel have shown that even on herbaceous Mediterranean grassland, primary production was not significantly reduced by heavy grazing compared with light grazing (Gutman *et al.*, 1990c). Even in desert ranges that have been heavily grazed for centuries, dramatic 'recovery' after relatively short periods of protection casts serious doubt on charges of lower production potential following intensive herbivory (Noy-Meir, 1990; Gillet & Le Houerou, 1991). In strongly seasonal pastures active growth takes place during a short period of the year, whilst animals have to graze the pastures for long periods, often all year round. Their numbers are regulated mainly by the forage availability and quality in the poor season (Perevolotsky, 1992). During the active growth pulse of the vegetation, daily defoliation rates per unit area are consequently much lower than pasture growth rates. When defoliation rates are much lower than growth rates, animal effects on the vegetation are small (Noy-Meir, 1975).

Long-term effects of grazing on primary production can be the result of a shift in species composition as a consequence of heavy grazing (Holechek *et al.*, 1989; Noy-Meir & Kaplan, 1990). Whilst primary production potential may not necessarily suffer, it is possible that the vegetation under grazing may include a greater proportion of unpalatable species that lead to lower secondary production. This could well have occurred in the Mediterranean region soon after the domestication of ruminants, because most of the dominant shrub and woodland species are unpalatable and have low preference indices, compared with selected pasture species (Papanastasis *et al.*, 1992). In addition, many have powerful biochemical and morphological protective mechanisms against herbivory (Perevolotsky *et al.*, 1991).

Secondary production. An important indicator of the effects of grazing on an ecosystem is the change in the number of animal units that the system can maintain (or the secondary production) over an extended period of time. On a country-wide basis, data of FAO (FAO 1947, 1950, 1971, 1988) on animal densities in the various countries around the Mediterranean Basin, indicate two main levels of variation: between groups of countries, and between years within the countries (Figs. 9.2A, 9.2B). In most countries (except Spain and Italy), animal densities have increased significantly over the past 50 years even where there was subsidized slaughter of goats to reduce grazing pressure (Papanastasis, 1986). Absolute numbers of small ruminants have been decreasing mainly in Spain, Italy and Albania, whilst in all other countries there has been an overall increase over the past (Figs. 9.2C, 9.2D).

There is, therefore, no indication in any country that the population of small-ruminants is heading to extinction or even massive reduction, whatever overgrazing there may be. Causes of diminishing numbers of animals are mainly socio-economic (Spain, Italy), whilst causes of increase most commonly include greater availability of supplementary feed (Tunisia, Algeria). National and EEC subsidies for small ruminant owners in hilly rural areas have also encouraged the increase of herds. In grazing studies on Mediterranean grassland, animal production per unit area has consistently been higher under heavy grazing (Crespo, 1985; Gutman *et al.*, 1990b), similar to

Fig. 9.2. Trends since 1940 in the density and relative flock size of small ruminants in some Mediterranean countries: A – Countries with lower densities (small ruminant units per hectare); B – Countries with higher densities; C – Countries with decreasing trends in flock size (normalized on flock size in each country in 1987); D – Countries with increasing trends. Figures based on data from FAO production Yearbooks for 1947, 1950, 1970 and 1987. Rangeland includes 'permanent pastures' and 'forested area'; small ruminant units are the sum of sheep and goat numbers cited.

the results recorded on some African rangelands (De Ridder & Wagenaar, 1986). In shrublands this would appear to be even more pronounced, because light grazing may lead to shrub encroachment, with a consequent reduction of access and of grazing value, at least for domestic livestock (Gutman *et al.*, 1990a).

Plasticity and resilience. Plasticity of an ecosystem can be defined as its capability to maintain the potential to change structurally, whilst resilience is the degree, manner and pace of recovery following disturbance (Westman, 1978). The appearance of Mediterranean brushlands and woodlands can change remarkably as a consequence of grazing or fire but, despite these differences in appearance, their floristic composition returns remarkably rapidly to the previous status even after so drastic a disturbance as fire (Poissonet *et al.*, 1978; Trabaud & Lepart, 1980, 1981; Malanson & Trabaud, 1987). Their conclusions 'reinforce the common perception of the garrigue as a very resilient plant community' and that the 'long evolutionary history of frequent disturbances' in the vegetation of the Mediterranean Basin may be the reason that it can 'better resist changes than . . . in the United States or Australia' (Malanson & Trabaud, 1987). Another example of the same characteristic is the relatively rapid development of oak woodlands in central and northern Israel after cessation of the intensive grazing that was practised in the area for thousands of years. Resilience, measured in terms of primary production, was found to be high also in Mediterranean grasslands (Noy-Meir & Walker, 1986).

Fire-proneness. Mediterranean ecosystems are notoriously fire-prone (Naveh, 1975). Even in Greece, where numbers of small ruminants have increased in the last decade, wild-fires are common (Papanastasis, 1988). Heavy grazing is one of the more effective means to reduce the fuel load or prevent its accumulation in Mediterranean shrublands (Bonnier, 1981; Etienne, 1989; Gutman *et al.*, 1990a; Perevolotsky, 1991b; Talamucci, 1991). Consequently, reduction of ruminant grazing on traditional rangelands in some countries around the Mediterranean Basin, has aggravated the fire hazard to such an extent that undergrazing has become a prominent management and political issue (Crespo, 1985; Perevolotsky, 1991a).

Is overgrazing a threat to Mediterranean Basin rangelands?

Put another way, do the negative feedbacks that limit the effects of heavy grazing on vegetation protect Mediterranean rangelands from degradation? Crawley (1983) states: 'Leaf feeding species like cows grazing on pasture grasses . . . typically have little effect on plant death rate. Their feeding is essentially parasitic rather than predatory: they reduce plant growth rate, but tend not to kill the plants.' He concludes (op. cit.): 'There is a fundamental asymmetry in plant-herbivore interactions. Plants have much more impact on the dynamics of herbivores than herbivores have on the dynamics of plants.' This could be related to the fact that on most rangelands, primary production is more than an order of magnitude higher than secondary production. The plant-animal relations in the Mediterranean Basin reflect this almost universal asymmetry. Fuentes and Etchegaray (1983) point out that the 'interaction between herbivores and their food sources, in which the expected benefits for herbivores are inversely related to those of plants, generates a minimax or saddlepoint solution . . . we neither see herbivores denude plants completely, nor herbivores completely excluded from the community; instead we observe an intermediate point of dynamic coexistence.' The high degree of resilience of the Mediterranean woodland and the persistence of small ruminant grazing in the Mediterranean Basin would seem to be an example of such a dynamic coexistence. This does not fit the 'fragile ecosystem threatened by overgrazing' paradigm that guides much of American range and vegetation management (Holechek *et al.*, 1989). On the contrary, only heavy grazing appears to be a viable principle for range management in much of the Mediterranean Basin; the problem is not so much vegetation and landscape stability as the consequent lower production level of the domestic livestock in a non-subsistence economy. In order to maintain acceptable levels of animal production under heavy grazing, suitable supplementary feeding and breeding schemes must be developed. This can lead to large increases in livestock numbers and, as a result, to the occurrence of severe, localized overgrazing and range degradation. As a rule, however, intensification aims at higher livestock production and that cannot be attained on severely depleted range. Consequently, when exogenous feed inputs are introduced and production goals are raised, the tendency is to substitute forage for range, and to reduce grazing pressure rather than to increase it. In addition, the extensive nature and remoteness of much of the rangeland is leading to increasing abandonment of traditional herding methods, thereby creating a new challenge: how should these lands be managed with much less, and more costly, labour? Whilst conservation is one answer, it does not solve other related problems – fire, unemployment, depopulation of rural areas, or loss of traditional products.

'The Mediterranean has a special claim to our interest . . . . It is the place where mankind's exploitation of the land began and where it has run its full cycle. What happened here during past millennia is, elsewhere on earth, just beginning' (Attenborough, 1987). Has the time not come to consider a new approach to grazing management of the rangelands of the Mediterranean Basin that takes into account their remarkable resilience as well as the multiple goals and constraints imposed on the system by an increasingly industrialized society?

References

Attenborough, D. 1987. The First Eden: The Mediterranean World and Man. Collins/BBC Books, London.

Berliner, R. 1970. The vegetation of the post-Eocene volcanic rocks in the Galilee. M.Sc. Thesis, Hebrew University of Jerusalem, Israel. (in Hebrew).

Bonnier, J. 1981. Rôle du pâturage dans la prévention des incendies de forêts. Foret Medit. 3: 71–72.

Caldwell, M.M., Meister, H.P., Tenhunen, J.D. & Lange, O.L. 1986. Canopy structure, light microclimate and leaf exchange of *Quercus coccifera* L. in a Portuguese macchia: measurements in different canopy layers and simulations with a canopy model. Trees 1: 25–41.

Crawley, M.J. 1983. Herbivory: The Dynamics of Animal-Plant Interactions. Studies in Ecology, Vol. 11. University of California Press, Berkeley and Los Angeles.

Crespo, D.G. 1985. Importance of grazing trials in determining the potential of rainfed Mediterranean pastures. FAO-European Cooperative Network on Pasture and Fodder Crop Production, Bulletin No. 4: 85–91.

Darwin, C. 1859. The Origin of the Species. John Murray, London.

De Ridder, N. & Wagenaar, K.T. 1986. A comparison between productivity of traditional livestock systems and ranching in eastern Botswana. In: Joss, P.J., Lynch, P.W. & Williams, O.B. (eds), Rangelands: A Resource Under Siege. Australian Academy of Science, Canberra, Australia, pp. 405–406.

Di Castri, F. & Mooney, H.A. 1973. Mediterranean-Type Ecosystems, Origin and Structure. Ecological Studies, Vol. 7. Springer-Verlag, Berlin.

Ellis, J.E. & Swift, D.M. 1988. Stability of African pastoral ecosystems: alternate paradigms and implications for development. J. Range Manage. 41: 450–459.

Etienne, M. 1989. Protection of Mediterranean forests against fire: an ecological approach for redevelopment. Paper presented at the Vth European Ecological Symposium (Sienna, Italy).

F.A.O. 1947. Yearbook of Food and Agriculture Statistics – 1947. FAO, Washington, DC.

F.A.O. 1950. Yearbook of production – 1950. Vol. IV(1), Rome, Italy.

F.A.O. 1971. Yearbook of production – 1970. Vol. 24, Rome, Italy.

F.A.O. 1988. Yearbook of production – 1987. Vol. 41, Rome, Italy.

Foran, B.D., Friedel, M.H., MacLead, N.D., Stafford-Smith, D.M. & Wilson, A.D. 1989. Policy Proposals for the Future of Australia's Rangelands. CSIRO National Rangelands Program, CSIRO, Lyneham, A.C.T., Australia.

Fuentes, E.R. & Etchegaray, J. 1983. Defoliation patterns in Matorral ecosystems. In: Kruger, F.J., Mitchell, D.T. & Jarvis, J.U.M. (eds), Mediterranean-Type Ecosystems: The Role of Nutrients. Ecological Studies 43, Springer-Verlag, Heidelberg, pp. 525–542.

Garrison, G.A. 1953. Effects of clipping on some range shrubs. J. Range Manage. 6: 309–317.

Gillet, H. & Le Houerou, H.N. 1991. Desert range exclosure and regeneration in the Arabian oryx reserve of Mahazed Assaid (Saudi Arabia). Proc. 4th International Rangeland Congress (Montpellier, France).

Gutman, M., Henkin, Z., Noy-Meir, I., Holzer, Z. & Seligman, N.G. 1990. Plant and animal responses to beef cattle grazing in Mediterranean oak scrub forest in Israel. Proc. 6th Meeting of the FAO-European Sub-Network on Mediterranean Pastures and Fodder Crops (Bari, Italy), pp. 191–196.

Gutman, M., Holzer, Z., Seligman, N.G. & Noy-Meir, I. 1990. Stocking density and production of a supplemented beef herd grazing yearlong on Mediterranean grassland. J. Range Manage. 43: 535–539.

Gutman, M., Seligman, N.G. & Noy-Meir, I. 1990. Herbage production of Mediterranean grassland under seasonal and yearlong grazing systems. J. Range Manage 43: 64–68.

Harper, J.L. 1969. The role of predation in vegetational diversity. Brookhaven Symp. Biol. 22: 48–62.

Holechek, J.L., Pieper, R.D. & Herbel, C.H. 1989. Range

Management Principles and Practices. Prentice-Hall, Englewood Cliffs, NJ, USA.

Le Houerou, H.N. 1981. Impact of man and his animals on Mediterranean vegetation. In: Di Castri F. & Specht, R.L. (eds), Ecosystems of the World, Vol. 11. Mediterranean-type Shrublands. Elsevier, Amsterdam, The Netherland, pp. 479–521.

Litav, M. 1965. Effects of soil type and competition on the occurrence of *Avena sterilis* L. in the Judean Hills (Israel). Isr. J. Bot. 14: 74–89.

Litav, M. 1967. Micro-environmental factors and species relationships in three batha associations in a foothill region of the Judean Hills. Isr. J. Bot. 16: 79–99.

Malanson, G.P. & Trabaud, L. 1987. Ordination analysis of components of resilience of *Quercus coccifera* garrigue. Ecology 68: 463–472.

Naveh, Z. 1971. The conservation of ecological diversity of Mediterranean ecosystems through ecological management. In: Duffey, E. & Watt, A.S. (eds), The Scientific Management of Animal and Plant Communities for Conservation. Blackwell, Oxford, UK, pp. 603–622

Naveh, Z. 1975. The evolutionary significance of fire in the Mediterranean region. Vegetatio 29: 199–208.

Naveh, Z. & Dan, J. 1973. The human degradation of Mediterranean landscapes in Israel. In: Di Castri, F. & Mooney, H.A. (eds), Mediterranean-Type Ecosystems: Origin and Structure. Ecological Studies, Vol. 7. Springer-Verlag, Berlin, pp. 373–390.

Naveh, Z. & Lieberman, A.S. 1984. Landscape Ecology. Springer Verlag, New York, NY.

Naveh, Z. & Whittaker, R.H. 1979. Measurements and relationships of plant species diversity in Mediterrranean shrublands and woodlands. In: Grassle, J.F., Patil, G.P., Smith, W.K. & Taille, C. (eds), Ecological Diversity in Theory and Practice. Statistical Ecology Series 6, International Cooperative Publishing House, MD, pp. 219–239.

Noy-Meir, I. 1975. Stability of grazing systems: an application of predator-prey graphs. J. Ecol. 63: 459–481.

Noy-Meir, I. 1978. Grazing and production in seasonal pastures: Analysis of a simple model. J. Appl. Ecol. 15: 809–835.

Noy-Meir, I. 1990. Response of two semiarid rangeland communities to protection from grazing. Isr. J. Bot. 39: 431–442.

Noy-Meir, I., Gutman, M. & Kaplan, Y. 1989. Response of Mediterranean grassland plants to grazing and protection. J. Ecol. 77: 290–310.

Noy-Meir, I. & Kaplan, Y. 1990. Effect of grazing on herbaceous vegetation and its implications for management of nature reserves. Research Report, Nature Reserves Authority, Jerusalem, Israel (in Hebrew).

Noy-Meir, I. & Walker, B.H. 1986. Stability and resilience in rangelands. In: Joss, P.J., Lynch, P.W. & Williams, O.B. (eds), Rangelands: A Resource Under Siege. Australian Academy of Science, Canberra, Australia, pp. 21–25.

Papageorgiou, N. 1979. Population energy relationships of the Agrimi (*Capra aegagrus cretica*) on Theodorou Island. Verlag Paul Parey, Hamburg & Berlin.

Papanastasis, V.P. 1986. Integrating goats into Mediterranean forests. Unasylva 38: 44–52.

Papanastasis, V.P. 1988. Rehabilitation and management of vegetation after wildfires in Maquis-type brushlands. Dasik Ereuna (Forestry Research) 10: 77–90 (in Greek, with an English summary).

Papanastasis, V.P., Nastis, A. & Tsiouvaras, C. 1992. Effects of goat grazing on species composition of variously treated *Quercus coccifera* L. ecosystems. In: Thanos, C.A. (ed), Plant-Animal Interactions in Mediterranean Type Ecosystems. Proc. VI International Conference on Mediterranean Climate Ecosystems. Athens, pp. 95–101.

Perevolotsky, A. 1991a. Animal-plant interactions: Contemporary progress and future challenges. Proceedings, 4th International Rangeland Congress (Montpellier, France).

Perevolotsky, A. 1991b. Rehabilitation of the black goat. Hassadeh 71: 619–622. (in Hebrew, with an English summary).

Perevolotsky, A. 1992. Goats or scapegoats – the overgrazing controversy in Piura, Peru. Small Rum. Res. 6: 199–215.

Perevolotsky, A., Haimov, Y. & Yonatan R. 1991. Feeding behavior of goats in Mediterranean woodland in Israel: an ecological-nutritional perspective. In: Thanos, C.A. (ed),. Plant-Animal Interactions in Mediterranean Type Ecosystems. Proc. VI International Conference on Mediterranean Climate Ecosystems. Athens, pp. 54–61.

Poissonet, P., Romane, F., Thiault, M. & Trabaud, L. 1978. Evolution d'une garrigue de *Quercus coccifera* L. soumise a divers traitements: quelques resultats des cinq premièrès années. Vegetatio 38: 135–142.

Rabinovich-Vin, A. & Orshan, G. 1974. Ecological studies on the vegetation of the Upper Galilee, Israel. 2. Factors determining the absence of batha and garrigue components on middle-Eocene strata. Isr. J. Bot. 23: 119.

Rozenzweig, D. 1972. Study of difference in effects of forest and other vegetation covers on water yield: Final Report. Research Report No. 33, Soil Conservation and Drainage Research Unit, Agricultural Research Organization, Volcani Center, Bet Dagan, Israel.

Safriel, U. 1991. Nature reserve management by grazing. Nature Reserve Authority, Jerusalem, Israel. Internal report, 14 pp. (in Hebrew).

Sakellariou, B. 1991. Niki Goulandris, a woman for all seasons. Chandris Hotels Magazine 7: 20–23.

Sandford, S. 1986. Information systems for range administration in developing countries with special reference to Africa. In: Joss, P.J., Lynch, P.W. & Williams, O.B. (eds), Rangelands: A Resource Under Siege. Australian Academy of Science, Canberra, Australia, pp. 509–512.

Seginer, I., Morin, Y. & Shachori, A. 1963. Experiments on runoff and erosion from the western slopes of Mount Carmel. Research report no. 8, Soil Conservation Service, Ministry of Agriculture, Tel Aviv, Israel. (in Hebrew).

Sela, Y. 1975. The recovery of the Mediterranean woodland – a problem in the conservation of wildlife. Teva Va'aretz 19: 81–84 (in Hebrew).

Shepherd, H.R. 1971. Effects of clipping on key browse species in southwestern Colorado. Division of Game, Fish and Parks, Techn. Publ. 28.

Shmida, A. 1981. Mediterranean vegetation in California and

Israel: Similarities and differences. Isr. J. Bot. 30: 105–123.

Specht, R.L., Grundy, R.I. & Specht, A. 1990. Species richness of plant communities: relationship with community growth and structure. Isr. J. Bot. 39: 465–480.

Specht, R.L. & Specht, A. 1989. Species richness of sclerophyll (heathy) communities in Australia – the influence of overstorey cover. Aust. J. Bot. 37: 337–350.

Stoddart, L.A., Smith, A.D. & Box, T.W. 1975. Range Management. McGraw-Hill Book Company, New York, NY.

Talamucci, P. 1991. Pascolo e bosco. L'Italia Forestale e Montana 46: 93–108.

Thirgood, J.V. 1981. Man and the Mediterranean Forest. Academic Press, New York, NY.

Thirgood, J.V. 1987. Cyprus: A Chronicle of its Forests, Land and People. University of British Columbia Press, Vancouver, Canada.

Tomaselli, R. 1977. The degradation of the Mediterranean maquis. Ambio 5: 356–362.

Trabaud, L. & Lepart, J. 1980. Diversity and stability in garrigue ecosystems after fire. Vegetatio 43: 49–57.

Trabaud, L. & Lepart, J. 1981. Changes in the floristic composition of a *Quercus coccifera* L. garrigue in relation to different fire regimes. Vegetatio 46: 105–116.

Tsiouvaras, C.N. 1987. Ecology and management of Kermes oak (*Quercus coccifera* L.) shrublands in Greece: A review. J. Range Manage. 40: 542–546.

Tsiouvaras, C.N., Noitsakis, B. & Papanastasis, V.P. 1986. Clipping intensity improves growth rate of Kermes oak twigs. Forest Ecol. and Manage. 15: 229–237.

Walter, H. 1968. Die Vegetation der Erde, Bd. 2: Die Gemassigten und Arktischen Zonen. G. Fischer, Jena, Germany.

Watt, A.S. 1981. A comparison of grazed and ungrazed grassland in East Anglian Breckland. J. Ecol. 69: 499–508.

Westman, W.E. 1978. Measuring the inertia and resilience of ecosystems. Bioscience 28: 705–710.

CHAPTER 10

Resource availability and herbivory in *Larrea tridentata*

PHILIP W. RUNDEL[1], M. RASOUL SHARIFI[1] and AZUCENA GONZALEZ-COLOMA[2]
[1]*Laboratory of Biomedical and Environmental Sciences and Department of Biology, University of California, Los Angeles, CA 90024, USA*
[2]*Instituto de Productos Naturale Organicos, CSIC, Avda. Astrofisico F. Sanchez 2, 38206 La Laguna, Tenerife, Canary Islands, Spain*

Key words: *Larrea tridentata*, acridid grasshopper, herbivory, NDGA, creosote bush

Abstract. *Larrea tridentata*, a widespread and ecologically dominant shrub throughout the warm desert regions of North America, is abundant in the Mediterranean-climate arid regions of California. As a long-lived evergreen shrub, *L. tridentata* can serve as a useful model system for improving our understanding of the interactions of resource availability and herbivore activity in evergreen sclerophyll species. Growing in an arid and nitrogen-limited environment, *L. tridentata* relies on carbon-based defences in the components of its external leaf resins. For herbivores which can tolerate these chemical defenses, however, variability of resource quality in leaf tissue significantly affects herbivore activity. The primary herbivores of *Larrea* are acridid grasshoppers with highly specialized feeding habits. The competitive interactions of these herbivores has led to the evolution of specific patterns of insect behaviour, particularly with respect to territoriality. Experimental changes in availability of water and/or nitrogen resources to host plants do not alter quantitative patterns of chemical defence, but significantly impact plant productivity and resource allocation. When highly-productive shrubs receiving water and nitrogen were compared with control shrubs, the biomass of leaves consumed by herbivores on productive shrubs was larger than on the control shrubs but herbivores nevertheless consumed a greater percentage of the leaves on control plants. Overall, however, herbivory is relatively low on *L. tridentata*.

Introduction

Larrea tridentata, creosote bush, is a widespread and ecologically dominant evergreen shrub throughout the arid regions of the southwestern United States and Mexico. While not a Mediterranean-climate species in the restricted sense, much of the range of *L. tridentata* in the Mojave and western Sonoran Deserts are arid Mediterranean-type ecosystems (MTEs) with winter precipitation predominant. As an evergreen desert shrub, *Larrea* potentially faces strong pressure from herbivores since leaves are present throughout the dry summer months when few other plant resources are available. Both its abundance and evergreen habit make *Larrea* a highly predictable resource (Barbour *et al.*, 1977 Schultz *et al.*, 1977). Despite this availability of *Larrea* leaves as a food resource, herbivory is relatively limited. Resinous compounds which coat the leaves of *Larrea* deter virtually all vertebrate herbivory. A small number of specialist arthropod herbivores are present, however, and these are notably successful in using *Larrea* as a resource (Schultz *et al.*, 1977).

Interactions between herbivores and evergreen plants in Mediterranean-climate regions represent critical biotic pressures for plant growth and survival (Mooney *et al.*, 1980; Lincoln & Mooney, 1984). While detailed studies are largely lacking, overall levels of herbivory on shrubs appear to be relatively small throughout the five MTEs of the world (Morrow, 1983). What factors control potential herbivory? Are low nutrient content and high sclerophylly in evergreen leaves important in restricting potential herbivory? Broad questions of this type cannot be answered simply (Mooney & Gulmon, 1982; Mooney *et al.*, 1983). The *Larrea* system, however, presents a useful model for study of the interactions of resource

M. Arianoutsou and R.H. Groves, Plant-Animal Interactions in Mediterranean-Type Ecosystems, 105–114, 1994.

availability to evergreen plants and activity by herbivores in MTEs. The ecology and physiology of *Larrea tridentata* have been studied for many years, and the population structure of herbivorous arthropods on *Larrea* is known in some detail.

In this chapter, we present an overview of studies on *Larrea tridentata* and the response of the plant and its herbivores to changes in the availability of water and nitrogen. We first review communities of phytophagous arthropods on *Larrea* and discuss orthopterans (grasshoppers) as specialized herbivores of particular significance. With this background, we describe our field experiments relating resource availability to plant growth and productivity in *Larrea tridentata* and discuss the impact of resource availability on arthropod herbivory. In doing this, we hope to point out the complexity involved in analyzing the biotic and abiotic factors as significant controls on herbivory in evergreen shrubs.

Larrea herbivores

Phytophagous arthropods on Larrea tridentata

Larrea tridentata (D.C.) Cov., because of its wide range and ecological dominance, has been the subject of a variety of studies of arthropod communities. These investigations have spanned sites in all of the major warm desert regions of North America, and comparable surveys have been carried out with other species of *Larrea* in the Monte Desert of South America. Populations of *L. tridentata* at Silver Bell, Arizona, in the Sonoran Desert supported 22 species of phytophagous insects, 14 defoliators and 8 sap feeders (Schultz *et al.*, 1977). These included six species of Orthoptera (grasshoppers), six Coleoptera (beetles), four Heteroptera (bugs), three species of larvae of Lepidoptera (moths and butterflies), and three membracids in the Homoptera. Eleven of these were obligate feeders on *Larrea* in this community. Of the 15 phytophagous insects which fed principally on leaf tissues, four were specialists on old leaves. The remaining species fed on both old and young leaves. Orthopterans at Silver Bell commonly hatched in early spring (February–April) and developed slowly into adults that reached peak activity and feeding in summer. In contrast, lepidopteran larvae reached peak activity earlier in March and April (Schultz *et al.*, 1977).

Detailed studies of seasonal arthropod abundance on *Larrea tridentata* were carried out during the International Biological Program (IBP) studies at Rock Valley, Nevada, in the Mojave Desert from 1971–75 (Mispagel, 1978; Rundel & Gibson, 1994). The two most important defoliators of *Larrea* at Rock Valley were the orthopteran *Bootettix argentatus* and the geometrid moth larvae *Semiothisa larreana. Bootettix* and *Semiothisa* together comprised the major part of all defoliators collected on *Larrea* at Rock Valley. As at Silver Bell, peak feeding activity of *Bootettix* and other grasshoppers occurred in mid to late summer compared to May and June for *Semiothisa*. Gelechid moth larvae and chrysonelid leaf beetles were also important herbivores on *Larrea* (Rundel & Gibson, 1994). Sap-feeding arthropods were abundant on *Larrea* at Rock Valley. These included treehoppers (Membracidae), leaf hoppers (Cicadellidae), mealy bugs (Pseudococcidae), plant bugs (Miridae), thrips (Thysanoptera) and mites (Acarina). Of these, the treehoppers and mealy bugs appeared to be the ecologically most important groups (Rundel & Gibson, 1994).

Lightfoot and Whitford (1987) investigated the significance of water and nitrogen availability in *Larrea tridentata* to densities of foliage insects in the Chihuahuan Desert of New Mexico. They found that populations of foliage arthropods were significantly increased on shrubs with improved nitrogen availability, but water supplementation alone did not have a significant effect. Control plants had a mean of 187 foliage arthropods within 28 species compared to 632 individuals in 33 species for shrubs fertilized with nitrogen. Sap-feeding insects formed the numerically largest group of arthropods in their study, but they did not provide data on the community structure of foliage insects.

Grasshoppers as herbivores of Larrea

Grasshoppers are important herbivores on *Larrea tridentata* throughout its range in North America. Like many grasshoppers, which are relatively specialized in their food habits (Bernays & Simpson, 1990), the predominant defoliators of *Larrea tridentata* are highly specific to *Larrea* and can thus be classified as monophagous. *Bootettix argentatus* is a strictly monophagous and ecologically successful herbivore of *Larrea tridentata* throughout the Mojave, Sonoran, and Chihuahuan Deserts (Otte & Joern, 1977). It will not eat other plant species even when faced with food deprivation (Chapman *et al.*, 1988). Individuals of *Bootettix* rest on the foliage of *Larrea* and are extremely cryptic. Territorial defence of host shrubs is relatively rare (Schowalter & Whitford, 1979), but mate guarding and associated intermale aggression is common, suggesting that male *B. argentatus* defend females rather than food resources (Greenfield & Shelly, 1990). *Bootettix* feeds principally on young leaf tissue (Schultz *et al.*, 1977).

Detailed population studies of *Bootettix argentatus* were carried out at Rock Valley during the IBP studies from 1971–75. Peak seasonal densities of *B. argentatus* ranged from less than 300 to nearly 1500 individuals · ha^{-1} (Rundel & Gibson, 1994). Year-to-year variations in densities of *B. argentatus* were related to a variety of abiotic and biotic factors. Favorable levels of soil moisture and soil temperature in the spring were important in breaking diapause and ensuring adequate survivorship of the egg stage. Nymph mortality depended strongly on air temperature and precipitation which affect levels of predation and parasitism on this vulnerable life stage. Years with relatively high precipitation at Rock Valley were associated with poor development of *Bootettix* because of predation by bombyliid flies of the genus *Mythiocoymia* (Rundel & Gibson, 1994). Seasonal population densities of *Bootettix* at Rock Valley were typically bimodal with a peak in June and July when nymphs appeared followed by a second peak about six weeks later in August as adults emerged (Fig. 10.1).

Ligurotettix coquilletti, a second widespread and ecologically important grasshopper feeding on *Larrea tridentata*, is oligophagous in the strict sense, but commonly feeds only on *Larrea* when this species is present. It has occasionally been found feeding on *Atriplex* or *Lycium* throughout the warm desert regions of North America (Rehn, 1923; Otte, 1981). Unlike *Bootettix argentatus*, *L. coquilletti* prefer mature leaf tissue. Males of *L. coquilletti* are highly sedentary and commonly spend virtually all of their adult life within a single *Larrea* shrub (Greenfield & Shelly, 1990). Males actively defend territories on these shrubs through overt aggression as well as passively by way of stridulation to warn off other males. The incessant clicking made by males in the morning and midday hours also serves to attract females. Not all *Larrea* shrubs exhibit similar densities of *Ligurotettix*, and shrubs with greater densities of males were not distinguishable by any structural trait (Greenfield *et al.*, 1987; Greenfield & Shelly, 1990). Leaf characteristics were important in distinguishing host plant quality for these herbivores. The implications of variations in host plant quality on territorial behavior in *L. coquilletti* has been described in detail in a series of papers by Greenfield and his colleagues (Greenfield *et al.*, 1987, 1989; Greenfield & Shelly, 1990; Shelly *et al.*, 1987).

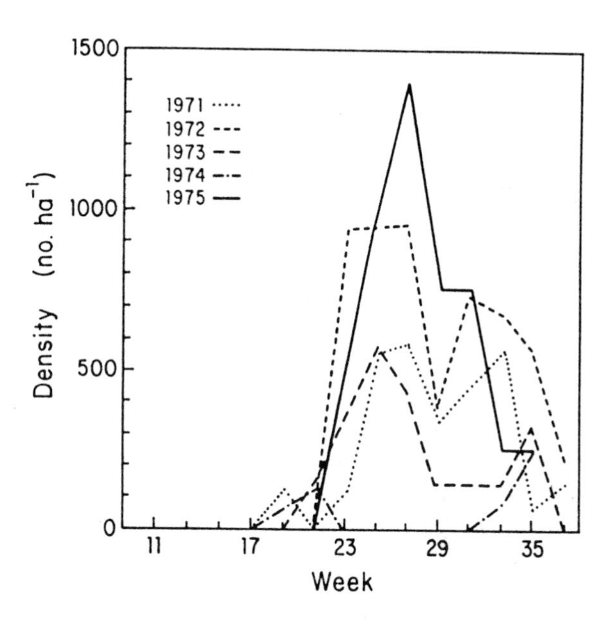

Fig. 10.1. Seasonal change in densities of *Bootettix argentatus* at IBP study site at Rock Valley, Nevada from 1971–75. From Rundel & Gibson (1993).

The most important herbivore on *Larrea cuneifolia* in the Monte Desert of Argentina is also a grasshopper, *Astroma quadrilobatum*. High densities of 50–100 individuals commonly occur on small to medium-sized *Larrea*, and may defoliate entire shrubs (Schultz *et al.*, 1977). *Astroma* prefers young leaves, but may eat leaves of all ages as well as flowers and flower buds.

Larrea *chemistry as a deterrent to herbivores*

The secondary chemistry of *Larrea* has been studied in detail (Seigler *et al.*, 1974; Mabry *et al.*, 1977; Rhoades, 1977b). The genus *Larrea* is well known for the abundant resins which coat the outer leaf surfaces. These resins are composed of a complex mixture of phenolics, saponins, terpenoids, and wax ester. Over 80% of the resin is composed of phenolic aglycones, with the major component nordihydroguiairetic acid (NDGA). NDGA is a strong antioxidant (Oliveto, 1972), with significant bactericidal, fungicidal and antiherbivore properties (Rhoades, 1977a,b; Greenfield *et al.*, 1987; Chapman *et al.*, 1988; Gonzalez-Coloma *et al.*, 1988).

Concentrations of leaf resins and associated chemicals decrease with leaf development as structural carbon is added to maturing leaves. Rhoades (1977b) reported that resins average 26.2% of the dry weight of young leaves of *Larrea tridentata*, but only 10.5% for mature leaves. NDGA content (expressed as catechols) were 12.5% of leaf dry weight for young leaves and 3.8% for old leaves. Such mean values, however, mask high levels of inter- and intrapopulational variation that are present in resin and NDGA content. Plant-to-plant variation in resin NDGA content varied greatly in a Sonoran Desert population of *Larrea tridentata* (Greenfield *et al.*, 1987, 1989). Concentrations of resin and NDGA, however, were relatively consistent from year-to-year in a given shrub (Greenfield, personal communication). Both genetic and abiotic factors may explain interpopulational variation in NDGA content. Downum *et al.* (1988) suggested that general decreases in NDGA content from north to south in the Sonoran Desert may be related to patterns of solar irradiance. Gonzalez-Coloma *et al.* (1988) showed that high ambient ozone concentrations could lead to breakdown of this antioxidant chemical. Shrub resource availability appears to have little to do with leaf resin and NDGA concentration.

NDGA plays a key role in food selection by specialist grasshoppers on *Larrea*. In experimental studies using differing concentrations of NDGA on sucrose-impregnated glass-fibre disks, Chapman *et al.* (1988) compared the feeding of three species of specialist grasshoppers. *Ligurotettix coquilletti* and *Cibolacris parviceps* were stimulated to feed by low concentrations of NDGA, but concentration of 0.5% or greater increasingly inhibited feeding (Fig. 10.2). Both of these species normally feed only on mature *Larrea* leaves (Schultz *et al.*, 1977) where NDGA concentrations are relatively low. The monophagous *Bootettix argentatus* was stimulated to feed at all concentrations of NDGA up to 20%.

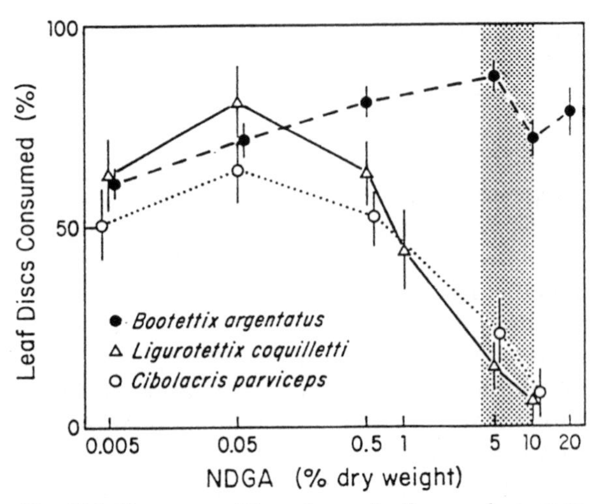

Fig. 10.2. Responses of three *Larrea*-feeding grasshoppers to experimental concentrations of NDGA presented on sucrose-impregnated glass fiber disks. The stippled area shows the range of NDGA occurring in mature *Larrea* leaves under field conditions (adapted from Chapman *et al.*, 1988).

Leaf resins, in general and NDGA, in particular, have been identified as major components of the antiherbivore system of *Larrea* (Rhoades, 1977a,b). In detailed studies with *Ligurotettix coquilletti*, however, NDGA concentration could explain only 10% of the variation in occupancy levels on individual *Larrea* (Greenfield *et al.*, 1989; Greenfield & Shelly, 1990). The significance of NDGA as an herbivore deterrent may lie not with this compound alone but together with other resin components. Gonzalez-Coloma *et al.* (1990) were able to demonstrate this effect with a generalist herbivore. *Larrea* resins contain complex mixtures of O-methylated flavones and flavonols as well as small amounts of acidic and basic components (Mabry *et al.*, 1977; Rhoades, 1977b), but little is known about the biological effects of these compounds.

Responses to resource availability

If herbivores respond largely to the quality of plant resources available, then changes in water and nitrogen availability to the plant could be expected to have significant impacts on the nature of leaf structure and protein content. In our study system, we have manipulated the availability of water and nitrogen resources of *Larrea tridentata* in a natural desert habitat. Our approach has been to assess how these changes in resource supply affected the quality and quantity of *Larrea* leaves available as well as the action of specialist herbivores.

Field manipulation treatments

Field manipulations of resource availability were carried out in a sandy wash woodland within the Living Desert Reserve near Palm Desert in the Sonoran Desert of California. *Larrea tridentata* is codominant in this community with *Cercidium floridum* (palo verde) and *Psorothamnus spinosus* (smoke tree). Other associated species are *Acacia greggii* (cat-claw acacia), *Chilopsis linearis* (desert willow), *Hymenoclea salsola* (cheesebush), and *Petalonyx thurberi* (sandpaper plant). The experimental design consisted of control plants plus three treatments – water, soil nitrogen, and water plus soil nitrogen. Three mature *Larrea* were included in each control and treatment group. Sufficient irrigation to bring soil water content to field capacity was applied monthly from April through November 1984 and March through October 1985 (Sharifi *et al.*, 1988). Soil nitrogen augmentations were added as NH_4NO_3 in a 4-m diameter circle around each shrub at a 5 g N m^{-2} after April irrigation of each year and 2.5 g N m^{-2} after the September irrigation.

Quantitative phenological measurements of plant growth were taken at regular intervals through the two years of experimental treatments. From these measurements, calculations could be made of leaf initiation, development, and senescence or loss to herbivory. Regression analyses of clipped branches allowed the calculations of productivity and changing biomass for each tissue component. Details of these phenological analyses were presented by Sharifi *et al.* (1988, 1990). Mature leaves were collected from four representative branches around each shrub. Total leaf resin content was determined by acetone extracts, and these extracts were analyzed for NDGA content by HPLC (Gonzalez-Coloma *et al.*, 1988). Leaf nitrogen content was determined using a Kjeldahl digest and autoanalyzer calorimetric analysis. Leaf protein content was measured using a modified Bradford method (Howard, 1987) calibrated against a RuBP carboxylase standard.

Treatment impacts on Larrea *growth*

Rates of shoot elongation and total leaf production were significantly higher in water and water plus nitrogen treatments than in control or nitrogen addition treatments (Fig. 10.3). In the treatments without water, both shoot elongation and leaf production had a bimodal pattern of peaks in spring and summer 1984. Unusual rains in July and August of this year provided the favourable summer conditions for growth. For the watered treatments, rates of stem elongation and total leaf production kept rising from spring until peaks were reached in later summer or early fall.

In the drier conditions of 1985, stem elongation and total leaf production was extremely small in the control and nitrogen treatments relative to the watered treatments (Fig. 10.3). Both of these water treatments again showed high rates of growth, peaking in September. In contrast to the previous year, however, rates of both stem elongation and total leaf production were significantly higher in the water plus nitrogen treatment than in the water treatment.

Comparative fruit production between treatments showed a very different pattern from that of stem elongation and leaf production. Control and nitrogen treatments without additional water had significantly higher rates of fruit production than either water treatment in 1984 and higher than the water treatment in 1985 (Fig. 10.3). Thus resource availability has a significant effect on carbon allocation. Treatments without addition of water had a mean allocation of clipping weight

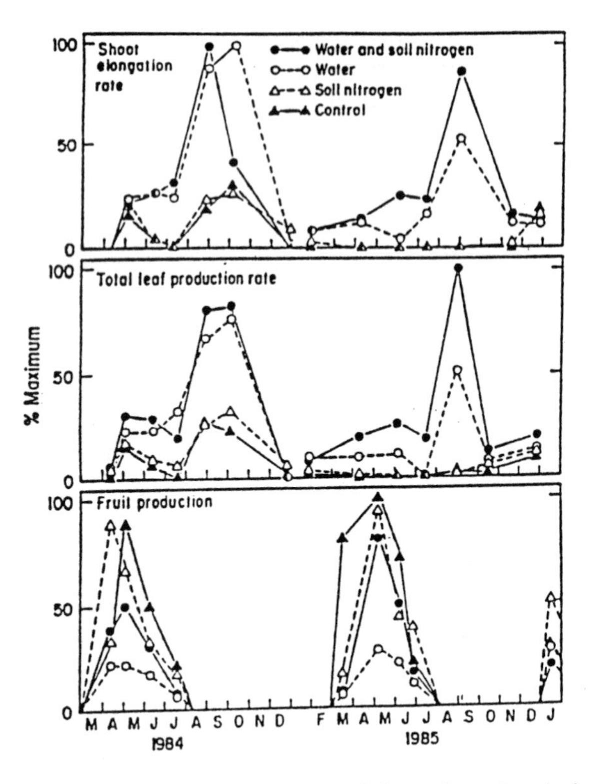

Fig. 10.3. Relative seasonal rates of shoot elongation, leaf production and fruit production of *Larrea tridentata* under four treatment regimes in 1984 and 1985 at Living Desert Reserve, California. Values presented are percentage of maximum rates observed in any treatment over the two years of study (from Sharifi *et al.*, 1990).

productions for fruit growth of 5.8 and 10.4% in 1985 and 1985 respectively, compared to a mean of 0.8 and 1.6% in these two years for the two watered treatments (Sharifi *et al.*, 1990).

Treatment impacts on leaf characteristics

Nitrogen contents of *Larrea* leaves showed significant patterns of difference between treatments. Mean leaf nitrogen content of control plants in 1984 was 12.8 mg g^{-1}, significantly lower than values for the other treatments (Table 10.1) In 1985, the pattern of impact was altered. Control, water and nitrogen treatments were not significantly different at 11 mg g^{-1}, while the mean nitrogen content of the water plus nitrogen treatment was 14.4 mg g^{-1}. The higher concentration in this latter treatment suggests that the effects of fertilization treatment were finally becoming significant. Protein content, however, was not significantly affected by treatment or year, with a mean value of about 90 mg g^{-1} overall.

Neither leaf resin content nor NDGA concentration were significantly affected by treatment or year, with resin averaging about 20% of dry weight and NDGA about 9%. It is important, however, to note the variability of these values among individual plants. In a wider survey of shrubs under natural conditions at the Boyd Deep Canyon Field Station nearby, resin contents varied from 2.1 to 8.5% and NDGA from 8.0 to 27.4 (Fig. 10.4).

Host plant quality and herbivory

The best available data on herbivory comes from *Larrea* populations at Living Desert Reserve and Deep Canyon where *Ligurotettix coquilletti* is the predominant herbivore. Host plant quality is clearly a primary factor in determining occupancy by *L. coquilletti*. Greenfield and Shelly (1990) have described how male individuals frequently move among bushes at the beginning of the season, but soon established fixed territories. The physical characteristics of *Larrea* shrubs (height, canopy area, stem density and nearest neighbour distance) were not significantly related to occupancy in their studies, but to genetic characteristics of host plant quality. Bushes heavily occupied in a given season were the first to be occupied by the territorial males, and maintained such pattems of occupancy from year-to-year (Fig. 10.5). Similarly, other shrubs consistently remained unoccupied each year.

Leaf nitrogen content also showed no direct correlation with occupancy by male *L. coquilletti*. Young leaves, the favoured food source of *Bootettix argentatus*, were consistently avoided in favor of mature leaves with lower nitrogen but reduced concentrations of resin and NDGA. Chemical analyses of leaf tissues from favoured shrubs established a significant correlation between occupancy and low concentration of hexane-extractable compounds (largely wax esters) and NDGA (Greenfield *et al.*, 1987, 1989; Greenfield & Shelly, 1990).

Laboratory feeding trials have supported the significance of these chemicals, particularly NDGA, as feeding deterrents. NDGA is a deter-

Table 10.1. Leaf nitrogen content (mg g^{-1} dry weight) and specific leaf weight (g m^{-2}) in four treatments of *Larrea tridentata* at Living Desert Reserve in the Sonoran Desert of California. Data are pooled samples of mature leaves from three individual shrubs in each treatment. Standard deviations of analyses are shown in parentheses

Treatment	Leaf nitrogen (mg g^{-1})		Specific leaf weight (g m^{-2})	
	August 1984	August 1985	July 1984	August 1985
Control	12.8 (1.1)	11.0 (1.9)	204	195
Nitrogen	14.7 (2.5)	11.5 (1.0)	218	205
Water	15.9 (0.9)	11.2 (1.1)	196	205
Water plus nitrogen	16.5 (0.6)	14.4 (1.1)	214	223

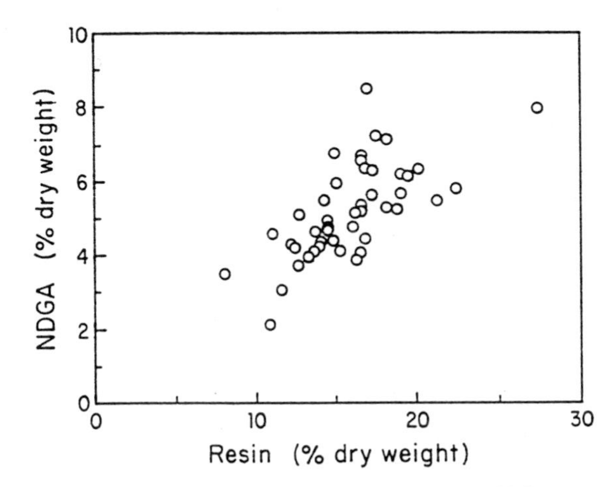

Fig. 10.4. Relationship between total resin and NDGA content of *Larrea tridentata* at the Boyd Deep Canyon Reserve, California. Each point represents a mean value for an individual shrub.

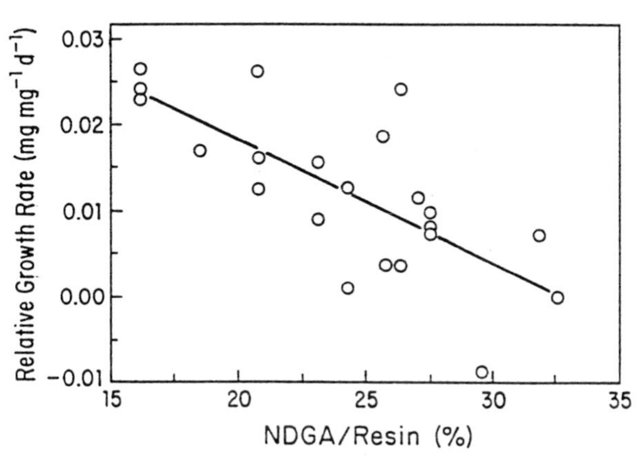

Fig. 10.6. Relationship between relative growth rate of female *Ligurotettix coqueilletti* and NDGA/resin ratio (the percentage of methanol-extraction resin comprised of NDGA) in the leaves of host shrubs of *Larrea tridentata* at the Boyd Deep Canyon Reserve, California. The least squares linear regression is significant at the 0.001 level, with $r = -0.72$ (adapted from Greenfield *et al.*, 1989).

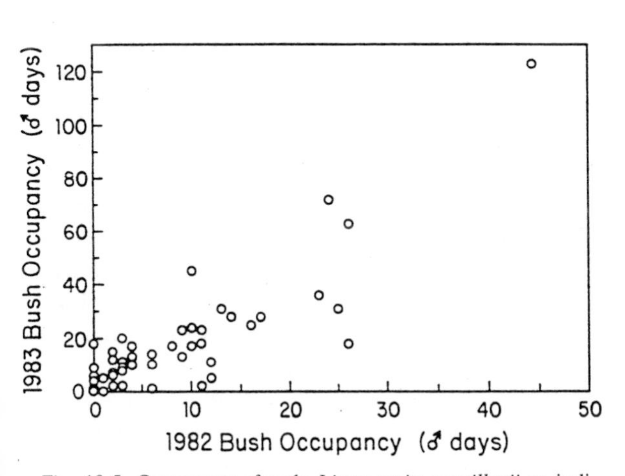

Fig. 10.5. Occupancy of male *Ligurotettix coquillettii* on individual shrubs of *Larrea tridentata* at the Boyd Deep Canyon Reserve California. The observation of one male on one census day (15 days in 1983 and 13 days in 1982) constituted one male day (adapted from Greenfield & Shelley, 1990).

rent to feeding for *L. coquilletti* at concentrations as low as 0.5%, much below the 3–10% range of concentrations typical of mature leaves (Fig. 10.2; Chapman *et al.*, 1988). Under field conditions, Greenfield *et al.* (1989) demonstrated that the relative growth rate of females on preferred shrubs was significantly higher than on shrubs with low or no occupancy. More specifically, relative growth rate of these females was inversely related to the NDGA content measured as a fraction of methanol-extractible resin component (Fig. 10.6).

Much remains unknown about the specific nature of chemical deterrents in host plant preference by *L. coquilletti*. Almost certainly, other compounds in addition to NDGA are important in this system. Further exploratory chemical work will be necessary to fully complete this story.

Direct measurements of total seasonal herbivory on *Larrea tridentata* by grasshoppers have not been made, but indirect studies have provided some interesting data. Using data from Rock Valley in the Mojave Desert, Mispagel (1978) calculated *Larrea* leaf consumption by *Bootettix argentatus* from his data on leaf energy equivalence, feeding efficiencies and *Bootettix* population size. For the years 1971–73, he estimated that *B. argentatus* consumed 1.0, 1.9 and 0.8% of the available leaf biomass of *Larrea*. This is a relatively small impact on plant production. Schultz *et al.* (1977), however, has reported significant defoliation of *Larrea cuneifolia* by orthoperans in Argentina.

In our own work, we have looked at the seasonal progression of herbivory in our resource manipulation experiments by counting the relative number of *Larrea* leaves in each treatment which show 25% or more of foliar area consumed by herbivores. This is a conservative measure since we could not sample leaves totally consumed because missing leaves from the previous measurement date could have been lost or abscised from a variety of causes.

Despite the conservative nature of these measurements, they provide an interesting perspective on seasonal patterns of herbivory. Relative herbivory levels, determined as the percent of the existing population of leaves showing 25% or more herbivory, peaked in summer in both 1984 and 1985 at the time of the year when *L. coquilletti* is most active as a herbivore (Fig. 10.7). Surprisingly, the highest levels of relative herbivory occurred in control plants in both years. Despite lower levels of foliar nitrogen than the water plus nitrogen treatments, relative consumption rates were 2–3 times higher in these control plants. Ecologically, however, these results are misleading because the pool sizes of leaves available for herbivores varied greatly between treatments and between years (Fig. 10.1). Relative herbivory levels of 4% for the water plus nitrogen treatment in 1984 involved a much larger biomass of leaves than the 8% herbivory on control plants. Similarly, generally higher rates of relative herbivory in 1985 can be attributed to lower net primary production and thus a smaller pool of resources available for *L. coquilletti*. Since we found no significant change in leaf resin or NDGA concentration with treatment or year, level of chemical defense is difficult to interpret as a variable in our experiments.

Conclusions

Bryant *et al.* (1983) have developed a model in which the carbon/nutrient balance of individual plants has a significant impact on plant allocation of resources to primary and secondary metabolism. Their model predicts that fertilization with limiting nutrients, usually nitrogen or phosphorus, will increase foliar concentration of such nutrients and thereby stimulate growth whilst at the same time decreasing leaf concentrations of carbon-based secondary metabolites. Their model is not fully supported by our experiments with *Larrea*. While nitrogen fertilization given with water did increase foliar nitrogen in the second year of treatment (consistent with results of Lightfoot and Whitford, 1987), it had no effect on the content of carbon-based defensive chemicals present as NDGA and total resin.

Abrahamson *et al.* (1988) have predicted from their studies with goldenrod that plants with strong genetic control of resistance to herbivory should show little change in their defensive chemistry with altered resource availability. Resource quality is of critical importance in understanding host plant selection by herbivores in *Larrea tridentata*. Consistent with predictions of resource availability and host plant defence against herbivores (Coley *et al.*, 1985), *Larrea* with its relatively slow growth rate, evergreen leaves and nitrogen-limited environment relies on carbon-based defences in its NDGA and other resin components. Nevertheless, the variability of resource quality for herbivores provides important ecological and evolutionary consequences for its herbivores.

Most herbivores avoid *Larrea* as a food resource, despite its apparency and availability throughout the year, because of toxic resins on the leaf surface. For those herbivores which have evolved mechanisms to detoxify these chemicals however, *Larrea* is a rich resource. Among these herbivores, in particular the grasshoppers, there

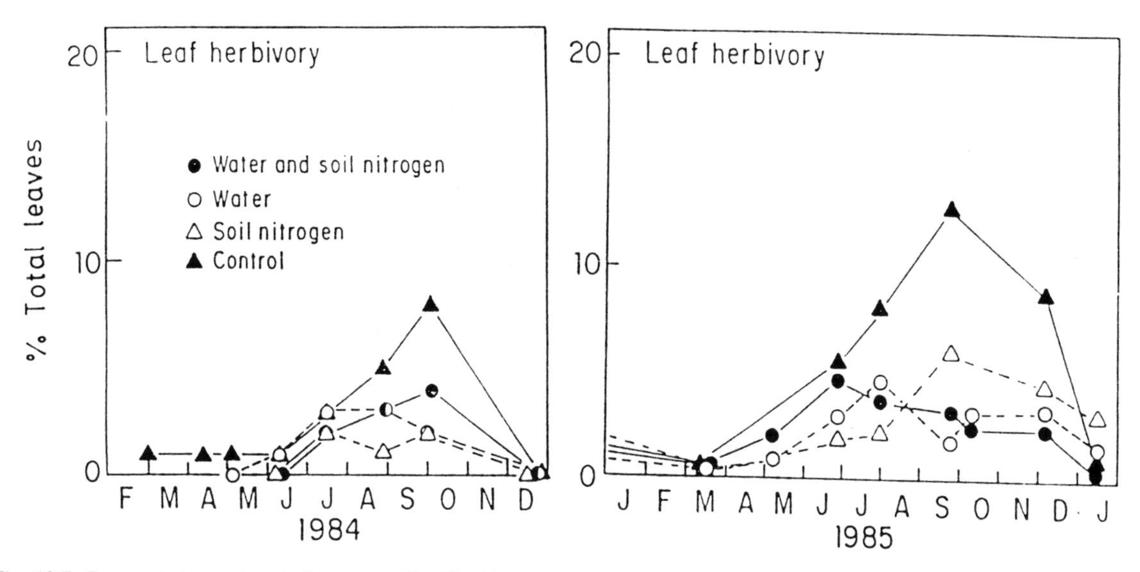

Fig. 10.7. Seasonal change in relative rates of leaf herbivory on *Larrea tridentata* under four treatment regimes in 1984 and 1985 at Living Desert Reserve, California. Data are the percentage of existing leaves showing 25% or more herbivory on the date of sample. Leaves totally consumed by herbivores since the previous sample date could not be determined and are thus not included.

is notable specialization in feeding patterns. *Ligurotettix coquilletti* avoids young leaves with their high resin content and feeds on mature leaves with lower resin but also lower protein content. *Bootettix argentatus*, in contrast, finds high resin content a stimulant to feeding (Chapman *et al.*, 1988) and consumes young foliage preferentially. Territoriality in these two species are the only examples known among acridid grasshoppers (Otte & Joern, 1975), thereby suggesting that their monophagy and unusual food resource are related to their patterns of behaviour (Greenfield & Shelly, 1990).

Herbivory on individual *Larrea* shrubs is highly variable, reflecting the variable quality of leaves for herbivores. This variation has a strong genetic component within populations of *Larrea* as reflected in consistent trends from year-to-year in patterns of grasshopper occupancy. Changes in resource availability for the host plant may have significant impacts on plant productivity, leaf biomass and nutrient resource availability. Leaf resin and NDGA content is not significantly affected by such changes, however. Because leaf chemistry controls herbivory much more strongly than does nutrient resource availability, increased plant productivity may not result in concomitant increases in relative amounts of herbivore impact on the resource. Overall, levels of herbivory on *Larrea tridentata* are relatively small.

References

Abrahamson, W.G., Anderson, S.S. & McCrea, E.D. 1988. Effects of manipulation of plant carbon nutrient balance on tall goldenrod resistance to a gall-making herbivore. Oecologia 77: 302–306.

Barbour, M.G., Cunningham, G., Oechel, W.C. & Bamberg, S.A. 1977. Growth and development, form and function, In: Hunziker, J.H. & DiFeo, D.R. (eds.) Creosote Bush: Biology and Chemistry of *Larrea* in New World Deserts. Dowden, Hutchinson and Ross, Stroudburg, Pennsylvania, pp. 48–91.

Bernays, E.A. & Simpson, S.J. 1990. Nutrition. In: Chapman, R.F. & Joern, A. (eds.) Biology of Grasshoppers. John Wiley and Sons, New York, pp. 105–127.

Bryant, J.P., Chapin, F.S. & Klein, D.R. 1983. Carbon/nutrient balance of boreal plants in relation to vertebrate herbivory. Oikos 40: 357–358.

Chapman, R.F., Bernays, E.A. & Wyatt, T. 1988. Chemical aspects of host-plant specificity in three *Larrea*-feeding grasshoppers. J. Chem. Ecol. 14: 561–579.

Coley, P.D., Bryant, J.P. & Chapin, F.S. 1985. Resource availability and plant antiherbivore defence. Science 230: 895–899.

Downum, K.R., Dole, J. & Rodriguez, E. 1988. Nordihydroguaiaretic acid: inter- and intrapopulational variation in the Sonoran Desert creosote bush (*Larrea tridentata*, Zygophyllaceae). Biochem. Syst. Ecol. 16: 551–555.

Gonzalez-Coloma, A., Wisdom, C.S. & Rundel, P.W. 1988. Ozone impact on the antioxidant nordihydroguaiaretic acid content in the external leaf resin of *Larrea tridentata*. Biochem. Syst. Ecol. 16: 59–64.

Gonzalez-Coloma, A., Wisdom, C.S. & Rundel, P.W. 1990. Compound interactions and effects of plant antioxidants in combination with carbaryl on performance of *Trichoplasia ni* (cabbage looper). J. Chem. Ecol. 16: 887–899.

Greenfield, M.D. & Shelly, T.E. 1990. Territory-based mating systems in desert grasshoppers: effects of host plant distribution and variation. In: Chapman, R.F. & Joern A. (eds.) Biology of Grasshoppers. John Wiley and Sons, New York, pp. 315–335.

Greenfield, M.D., Shelly, T.E. & Downum, K.R. 1987. Variation in host-plant quality: implications for territoriality in a desert grasshopper. Ecology 68: 828–838.

Greenfield, M.D. Shelly, T.E. & Gonzalez-Coloma, A. 1989. Territory selection in a desert grasshopper: the maximization of conversion efficiency on a chemically defended shrub. J. Anim. Ecol. 58: 761–771.

Howard J.J. 1987. Leafcutting ant diet selection: the role of nutrients, water and secondary chemistry. Ecology 68: 503–515.

Lightfoot, D.C. & Whitford, W.G. 1987. Variation in insect densities on desert creosote bush: Is nitrogen a factor? Ecology 68: 547–557.

Lincoln, D.E. & Mooney, H.A. 1984. Herbivory on *Diplacus aurantiacus* shrubs in sun and shade. Oecologia 64: 173–176.

Mabry, T.J., DiFeo, D.R., Sakakibara, M., Bohnstedt, C.F. & Siegler, D.S. 1977. The natural products chemistry of *Larrea*. In: Hunziker, J.H. and DiFeo, D.F. (eds.) Creosote Bush: Biology and Chemistry of *Larrea* in New World Deserts. Dowden, Hutchinson and Ross, Stroudsburg, Pennsylvania, pp. 115–134.

Mispagel, M.E. 1978. The ecology and bioenergetics of the acridid grasshopper, *Bootettix punctatus* on creosote bush, *Larrea tridentata*, in the northern Mojave Desert. Ecology 59: 779–788.

Mooney, H.A., Ehrlich, P.R., Lincoln, D.E. & Williams, K.S. 1980. Environmental controls on the seasonality of a drought deciduous shrub, *Diplacus aurantiacus* and its predator, the checkerspot butterfly, *Euphydryas chalcedona*. Oecologia 45: 143–146.

Mooney, H.A. & Gulmon, S.L. 1982. Constraints on leaf structure and function in reference to herbivory. Bioscience 32: 198–206.

Mooney, H.A., Gulmon, S.L. & Johnson, N.D. 1983. Physiological constraints on plant chemical defences. In: Hedin, P.A. (ed.) Plant Resistance to Insects. Amer. Chem. Soc., Washington, D.C, pp. 21–36.

Morrow, P.A. 1983. The role of sclerophyllous leaves in determining insect grazing damage. In: Kruger, F.J., Mitchell, D.T. & Jarvis, J.U.M. (eds.) Mediterranean-type Ecosystems: The Role of Nutrients. Springer-Verlag, Heidelberg, pp. 509–524.

Oliveto, E.P. 1972. Nordihydroguaiaretic acid: a naturally occurring antioxidant. Chem. Ind. 17: 677–679.

Otte, D. 1981. The North American Grasshoppers. I. Acrididae, Gomphocerinae and Acridinae. Harvard Univ. Press, Cambridge.

Otte D. & Joern, A. 1975. Insect territoriality and its evolution: Population studies of desert grasshoppers on creosote bushes. J. Anim. Ecol. 44:29–54.

Otte, D. & Joern, A. 1977. On feeding patterns in desert grasshoppers and the evolution of specialized diets. Proc. Acad. Nat. Sci. Philadelphia 128: 89–126.

Rehn, J.A.G. 1923. North American Acrididae (Orthoptera). 3. A study of the Ligurotettigi. Trans. Amer. Entomol. Soc. 49: 43–92.

Rhoades, D.F. 1977a. Integrated antiherbivore, antidesiccant and ultraviolet screening properties of creosote resin. Biochem. Syst. Ecol. 5: 281–290.

Rhoades, D.F. I977b. The antiherbivore chemistry of *Larrea*. In: Mabry, T.J., Hunziker, J.H. & DiFeo, D.R. (eds.) Creosote Bush: Biology and Chemistry of *Larrea* in New World Deserts. Dowden, Hutchinson and Ross, Stroudsburg, Pennsylvania, pp. 135–175.

Rundel, P.W. & Gibson, A. 1993. Ecological Process and Communities in a Mojave Desert Ecosystem. Cambridge University Press, Cambridge (in press).

Schowalter, T.D. & Whitford, W.G. 1979. Territoriality behavior of *Bootettix argentatus* Bruner (Orthoptera, Acridae). Amer. Midl. Nat. 102: 182–184.

Schultz, J.C., Otte, D. & Enders, F. 1977. *Larrea* as a habitat component for desert arthropods. In: Mabry, T.J., Hunziker, J.H. & DiFeo, D.R. (eds.) Creosote Bush: Biology and Chemistry of *Larrea* in New World Deserts. Dowden Hutchinson and Ross, Stroudsburg, Pennsylvania, pp. 176–208.

Seigler, D.S. Jakupcak, J. & Mabry, T.J. 1974. Wax esters from *Larrea divaricata* Cav. Phytochemistry 13: 983–986.

Sharifi, M.R., Meinzer, F.C., Nilsen, E.T., Rundel, P.W., Virginia, R.A., Jarrell, W.M. & Herman, D.J. 1988. Effect of resource manipulation on the quantitative phenology of *Larrea tridentata*. Amer. J. Bot. 75: 1163–1174.

Sharifi, M.R., Meinzer, F.C., Rundel, P.W. & Nilsen, E.T. 1990. Effect of manipulating soil water and nitrogen regimes or clipping production and water relations of creosote bush. In: Symposium on Cheatgrass Invasion, Shrub Die-off and other Aspects of Shrub Biology and Management. USDA Forest Service. Gen. Tech. Rep. INT-276, pp. 245–249.

Shelly, T.E., Greenfield, M.D. & Downum, K.R. 1987. Variation in host plant quality: influences on the mating system of a desert grasshopper. Anim. Behav. 35: 1200–1209.

CHAPTER 11

Effects of insect herbivory on plant architecture

ROSANNA GINOCCHIO AND GLORIA MONTENEGRO
Departamento de Ecología, Facultad de Ciencias Biológicas, Pontificia Universidad Católica de Chile, Casilla 114-D, Santiago, Chile

Key words: metameric architecture, convergence, herbivory, Mediterranean ecosystems

Abstract. In plants of Mediterranean-type ecosystems, the diversity and distribution of construction units or modules of species may lead to an enhanced understanding of specific types of metameric architecture and thus of canopy construction. An analysis of 160 species of plants from ecosystems of central Chile, France, Israel and South Africa resulted in the identification of six basic units of similar morphology and structure, which originate from development of either an apical or a lateral bud.

The activities of either gall-forming insects, that infect either vegetative or reproductive buds, or of bud-eating insects, may alter plant architecture, changes which also depend on plant morphology, spatial arrangement of plant parts and the timing of module loss by herbivory. The relationship between two gall-forming insects to the growth and development of the Chilean evergreen shrub *Colliguaja odorifera* was studied and the results are presented in this chapter.

Introduction

Each plant is organized as a population of units, metamers or modules (Lowell & Lowell, 1985) which are iterated by one or many growing points (Hallé *et al.*, 1978; White, 1979; Hallé, 1986). This serial repetition of plant organs by meristem activity has been called metameric growth (Hallé, 1986). Apical meristems create building blocks or modules which in turn develop an orderly arrangement of structures or units that generate a metameric architecture (Küppers, 1989). A plant grows by accumulating basic units and its architecture may be defined in terms of the types of modules it bears, their location and the times of their formation and death (Harper & Bell, 1979).

An understanding of the developmental sequence that leads to a specific type of architecture demands recognition of the building blocks or modules. Different module types have been distinguished at several levels of structure, such as a bud (Harper & Bell, 1979; Watkinson & White, 1986), a node with an axillary bud and leaf (White, 1984), a shoot with a terminal inflorescence (Prevost, 1966; Hallé, 1978) and a branch system or a ramet (Harper & Bell, 1979). In this chapter we follow the definition of a module by Hallé & Oldeman (1970) as a shoot unit resulting from the development of an axillary or apical bud, since every shoot on a plant commenced as a bud some time. The bud, then, may be seen as an unextended module that may live or die, remain dormant or grow into a shoot. Although meristem potential, position and fate determine plant morphology, the demography of a population of meristems has scarcely been explored; a considerable expansion of research on this topic may be anticipated in future.

Metameric architecture is closely related to the phylogenetic history of each species and to environmental conditions, such as temperature, light, water and nutrients. Changes in the metameric architecture of any given plant may arise because of the occurrence of different proportions of the developed modules and their spatial arrangement, due to changes in resource levels to plants, as well as the effects of factors that remove biomass, such as herbivory and fire (Küppers, 1989; Ginocchio & Montenegro, 1992; Ginocchio *et al.*, 1993).

By monitoring throughout the year the dyna-

M. Arianoutsou and R.H. Groves, Plant-Animal Interactions in Mediterranean-Type Ecosystems, 115–122, 1994.

mics of the bud bank, i.e. the dormancy, potential development, growth and fate of individual buds and the organs that senesce periodically in relation to the type of organs that remain on the plant, it was possible to construct growth patterns for each dominant species in the plant communities in Mediterranean-climate regions of central Chile (Montenegro *et al.*, 1989), France (Floret *et al.*, 1989), Israel (Orshan, 1989a) and South Africa (Le Roux *et al.*, 1989). The dynamic sequence of the events described allows us to identify and to determine the relative importance of the different construction units or modules responsible for the specific type of architecture that are representative of shrubs in Mediterranean-type ecosystems. We thus attempted to describe convergence patterns at the community level. This basic information has also been used to predict changes in plant architecture after bud removal by insect herbivores.

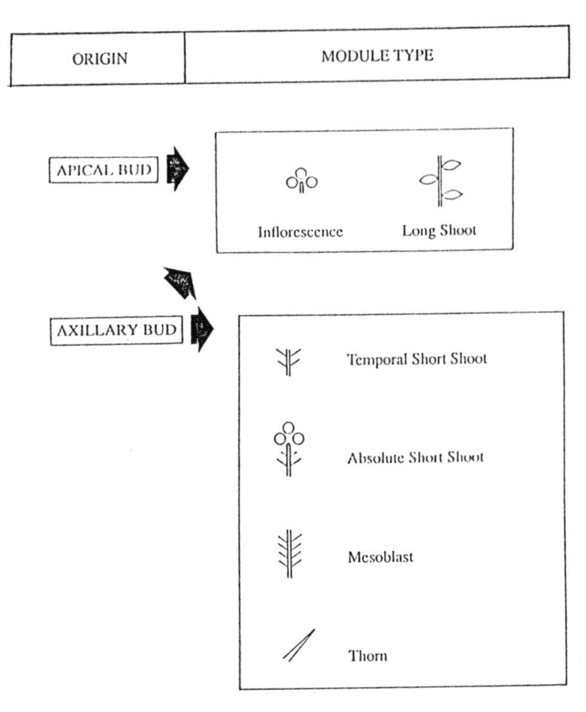

Fig. 11.1. Type of construction units developed from an apical or axillary bud described for Mediterranean-type ecosystem species.

Basic modules detected in species of Mediterranean-type ecosystems

The basic modules were detected by analyzing the life cycles of species dominant in Mediterranean-type communities of central Chile, France, Israel and South Africa. The annual cycles analyzed in this chapter were based partly on information published by Hoffman (1972), Hoffman & Hoffman (1976), Montenegro *et al.* (1980, 1988, 1989), Floret *et al.* (1989), Le Roux *et al.* (1989) and Orshan (1989a). The cycles were all obtained using similar methodologies. The seasonal progression of phenomorphological events was determined by random selection of ten individuals of each species. Ten shoots per shrub were marked at the level of the last leaf formed during the previous growth period (Montenegro *et al.*, 1979; Orshan, 1989b). The position of renewal buds, the flushing period, and the type of shoot arising from each bud were recorded. Observations were carried out once a week during the growth period and monthly for the rest of the year.

Despite the diversity of mediterranean shrub species in terms of growth form and leaf seasonality, the analysis revealed the existence of similar basic morphological units that are responsible for the construction of the canopy. Analysis of 160 species led to the identification of six basic structural units (Fig. 11.1) that originated from development of either an apical or a lateral bud. Apical buds were found to differentiate into an infloresecence or a vegetative long shoot which has the potential to continue growing vegetatively in the next season from its apical meristem. Axillary buds showed a greater ontogenetic plasticity; they were able to differentiate into any of the six different modules found. Besides the two units already described, a temporal short shoot may develop, which may or may not flower from its axillary buds. These temporal short shoots will produce a long shoot in the next growing season. Lateral buds may also give rise to an absolute short shoot limited by the transformation of the apical meristem into an inflorescence. Lateral buds may also originate a mesoblast that morphologically corresponds to a short shoot that can grow another short shoot in the next season. Finally, lateral buds may develop into a short shoot that lignifies and differentiates into a thorn. The modules described are combined in different spa-

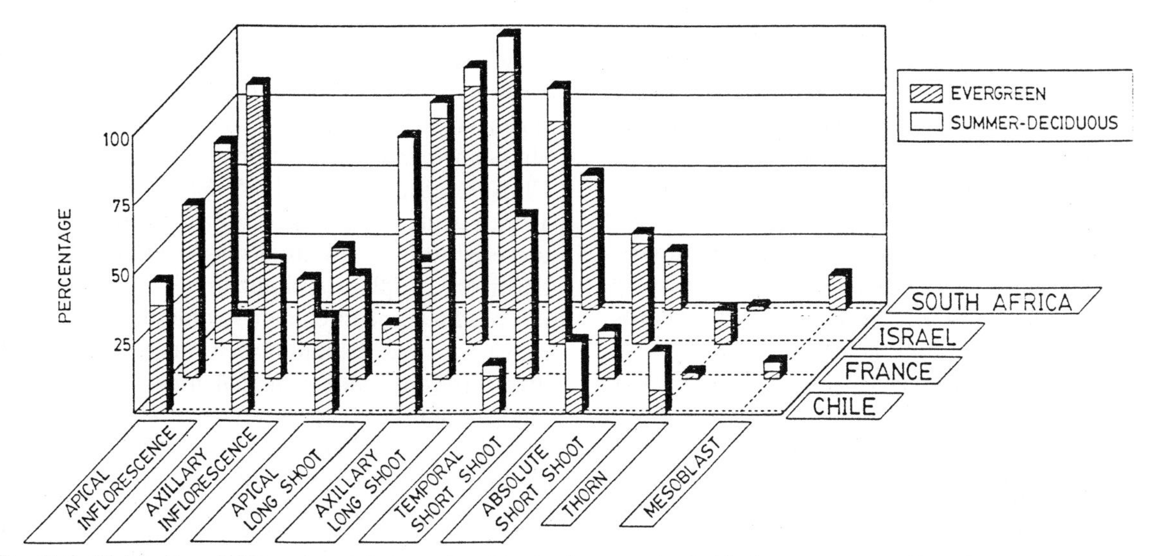

Fig. 11.2. Distribution of different module types found in dominant species in Mediterranean vegetation of Chile, France, Israel and South Africa.

tial arrangements to give a specific shape to the canopy (Ginocchio & Montenegro, 1992).

The frequencies at which each module type is represented in the different communities studied are shown in Figure 11.2. Frequency data, tested for independence by statistical analysis of multidimensional contingency tables (Everitt, 1977), showed similar distributions between Chile and France, and also between Israel and South Africa ($P < 0.001$). Species of the first pair of countries appear to develop more apical long shoots, whilst those from Israel and South Africa develop more apical inflorescences. These results relate to differences found in the abundance of chamaephytes and phanerophytes between the four countries (see Figure 11.3).

The growth habit of axillary long shoots, with a rapid expansion of internodes followed by a dormant period, was observed to predominate in both evergreen and summer-deciduous species (analysis of independence by multidimensional contingency tables, $P < 0.001$), thereby allowing a wide range of vertical stratification. Since one of the most important environmental resources for plants is light intensity, the occupation of a higher stratum above the ground may give a competitive advantage over shorter plants (Lowell & Lowell, 1985) as a result of gaining access to more available resources (Franco, 1986).

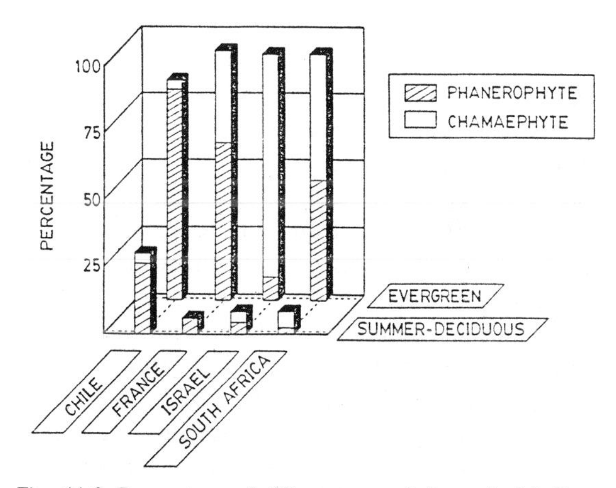

Fig. 11.3. Percentage of different growth forms in Mediterranean vegetation of Chile ($n = 23$), France ($n = 34$), Israel ($n = 25$) and South Africa ($n = 78$).

Short shoots obviously permit a lower plant height since little or no internode elongation occurs during the developmental process. Short shoots are important, however, especially in summer-deciduous species ($P < 0.05$) as structures that permit a rapid replacement of leaf area in order to increase photosynthesis with less resource and effort expended in internode formation. Summer-deciduous species in Mediterranean-type ecosystems shed their absolute short shoots and leaves of their long shoots at the end of the vegetative and reproductive periods respectively, thereby lowering the costs to maintain these structures (Orians & Solbrig, 1977; Oechel *et al.*, 1981).

A significant number of evergreen species develop both long and temporal short shoots showing greater plasticity under Mediterranean-type climatic conditions, since these plants gradually shed their long shoot leaves in summer, thereby reducing the transpiring area; their short shoots with markedly smaller leaves remain as the only green component of the plant. Temporal short shoots were also observed in summer-deciduous species of Chile, Israel and South Africa, but they differ from the ones developed in evergreens in their time of formation. In summer-deciduous species, temporal short shoots are formed at the beginning of the growing period, turning into long shoots in the same season, whilst in evergreens they are formed at the end of the growing season and bear the renewal buds.

Relationship between insect herbivory and plant architecture

Although herbivores frequently consume only a small proportion of the net primary production of terrestrial ecosystems (Golley, 1973; Chew, 1974; Owen & Weigert, 1976), they may have important effects on ecosystem structure and function. Several authors have shown that herbivores have effects on plant establishment (Gashwiller, 1970; Louda, 1983), growth (Morrow & La Marche, 1978; Hilbert *et al.*, 1981; McNaughton, 1983) and reproductive success (Janzen, 1969; Chew & Chew, 1970; Louda, 1983). Herbivores may also have a substantial effect on plant form (McNaughton, 1976, 1979, 1983; Harper, 1977; Owen, 1980). Herbivores usually remove only parts of the plant, leaving other plant parts capable of regeneration through the iteration of new modules (Watkinson & White, 1986). Although the development of dormant lateral buds may enable repair of damage (Harper, 1977; Hardwick, 1986), the plant undergoes a change in form because of an increase in branching and a more dense and bushy geometry results (McNaughton, 1984).

Even though the results of various studies have shown that herbivores (both browsers and grazers) may play an important role in Mediterranean-type ecosystems, by altering both the abundance and composition of species (see, e.g., Christensen & Muller, 1975; Mills, 1983; Quinn, 1986; Fuentes *et al.*, 1987; Frazer & Davis, 1988), there is little information on the effect of insect herbivory on plant architecture. The few studies in Mediterranean-climate communities relate only to shrub defoliation by chewing insects which give different values for leaf area lost. For the Chilean matorral the average leaf area lost is between 8 and 10% (Fuentes *et al.*, 1981), whereas comparable data for South African fynbos and Australian mallee are 2 and almost 30% respectively (Morrow, 1983). Although insect defoliators remove photosynthetic tissue, thereby leading to a consequent loss in productivity, they do not induce changes in plant architecture unless they damage or remove the growing points or buds. Gall-forming insects which infect vegetative or reproductive buds, and bud-eaters, which directly affect plant architecture, have rarely been described for Mediterranean-type plant communities. They are probably important components of the systems, as has been observed for matorral in central Chile (Table 11.1).

How can gall-formers and bud-eaters change plant architecture? In an attempt to describe in a dynamic way the effect of insect herbivory on plant architecture we studied the relationship of two gall-forming insects to the evergreen shrub *Colliguaja odorifera* (Euphorbiaceae). *C. odorifera* is a dominant shrub in the matorral vegetation with a relative cover of between 5 and 10%; it is not defoliated by herbivorous insects (Montenegro *et al.*, 1980; Walkowiak *et al.*, 1984). Reproductive and vegetative buds are infected by two different gall-forming insects *Torymus laetus* (Torymidae) and *Exurus colliguayae* (Eulophidae), both of which belong to the Superfamily Chalcidoidae (Hymenoptera). Even though both species may attack reproductive and vegetative buds on the same shrub (Martinez *et al.*, 1992), *E. colliguayae* tends to be more specific to flower buds (see Table 1) than *T. laetus*, and vice versa ($P < 0.001$, Chi square test). Besides differing in the type of bud targeted, the two

Table 11.1. Insect attack in relation to their plant reactive sites in shrub species of the matorral in central Chile

Plant host	Insect	Reactive site	Reference
Baccharis linearis	*Rachiptera limbata* (Diptera)	Apical bud of long shoots	Aljaro *et al.* (1984)
	(Hymenoptera)	Apical bud of temporal and absolute short shoots	Ginocchio & Montenegro, unpublished data
Bahia ambrosioides	Gryllidae spp (Orthoptera)	Leaves	Solervicens & Elgueta (1989)
Colliguaja odorifera	*Exurus colliguayae* (Hymenoptera)	Axillary vegetative buds & Apical reproductive buds	Martínez, Montenegro & Elgueta (1991)
	Torymus laetus (Hymenoptera)		
Cryptocarya alba	(Lepidoptera)	Leaves	Etchegaray & Fuentes (1980)
	Chrysomelidae (Coleoptera)		
Fuchsia lycioides	*Conomyrma chilensis* (Hymenoptera)		Solervicens & Elgueta (1989)
Kageneckia oblonga	(Lepidoptera)		Etchegaray & Fuentes (1980)
	Chrysomelidae (Coleoptera)		
Lithrea caustica	*Procalus lenzi* (Coleoptera)		Grez (1988)
	Procalus malaisei (Coleoptera)		
Muehlenbeckia hastulata	*Macrophalia* sp (Lepidoptera)		Fuentes *et al.* (1981)
	Chrysomelidae (Coleoptera)		
Peumus boldus	Unidentified	Vegetative and reproductive apical buds	Ginocchio & Montenegro, unpublished data
Quillaja saponaria	Chrysomelidae (Coleoptera)	Leaves	Etchegaray & Fuentes (1980)
Schinus polygamus	*Tainarys sordidata* (Psyllidae)	Vegetative axillary and apical bud and leaves	Flores, Mujica, Gómez & Montenegro (1987)
	(Hymenoptera)	Vegetative axillary buds	
Talguenea quinquinervia	(Lepidoptera)	Leaves	Etchegaray & Fuentes (1980)
	Chrysomelidae (Coleoptera)		
Trevoa trinervis	Chrysonmelidae (Coleoptera)		

species also differ in the intensity of attack (Table 11.2).

Plants of *Colliguaja odorifera* start activity in July by flowering, which occurs from pre-formed apical reproductive buds. Vegetative growth starts later with branch development from the uppermost axillary buds, thereby giving a typical Y-shaped branching pattern (Hoffman & Hoffman, 1976; Montenegro *et al.*, 1988, 1989). After growth finishes the apical bud differentiates into a flower bud that remains dormant until next growing season. This monoecious shrub develops an apical male inflorescence with two female flowers at the base.

Insect oviposition takes place in the resting period, i.e. in early summer. When the apical male inflorescence is infected, the sub-apical female flower buds do not develop into flowers, probably because of nutrient depletion induced by the growing gall (Abrahamson & Weis, 1987). Infected vegetative buds do not develop branches and thus the typical Y-shaped form of the branching pattern is changed. This latter type of infection also implies a time-lagged castration, since

Table 11.2. Frequency of gall-former insect species related to each gall type found in *Colliguaja odorifera* and frequency of each gall type per shrub. (E. Martinez, unpublished data)

	Frequency of gall-inducing insect (n=173)		Mean counts (n=22) and variance (in brackets) of gall types per area of shrub (0.25 m^2)
	Exurus colliguayae	*Torymus laetus*	
Galls in aments	113	17	3.6 (6.01)
Galls in vegetative buds	8	31	5.4 (4.71)

Table 11.3. Fruit production in infected and control *Colliguaja odorifera* shrubs.

	Mean counts and variance (in brackets) of fruits per rate of shrubs (0.25 m^2)		Decrease in the fruit setting between 1987 and 1988
	1987	1988	
Uninfected shrubs (n=11)	6.4 (11.23)	1.3 (1.89)	79.7%
Infected shrubs (n=22)	9.2 (15.46)	7.9 (17.55)	14.1%

Data from Martínez, Montenegro & Elgueta (1992).

the new branches are responsible for formation of flower buds at the apex. In this sense, attack by *T. laetus* on vegetative buds not only affects plant architecture but also the potential for seed production of *E. colliguayae* by directly affecting the flower buds.

Infection of the male flowers by *E. colliguayae* induces suppression of fruit growth. Since the reproductive budget to female structures is far greater than that imposed by male structures, because of the cost of producing seeds (Putwain & Harper, 1972), the resources available can then be allocated to vegetative growth. As a perennial polycarp shrub, *Colliguaja odorifera* shows an inverse correlation between reproduction and growth (Harper, 1977).

A change in pattern was observed in plants of *C. odorifera* that develop over the same period of infection a greater number of branches arising from lateral buds. Production of fruits was greater on these plants as a result of the large number of branches that bear flower buds. A comparison of the mean number of fruits produced between 1987 and 1988 showed a reduction in fruit production in the second year (Table 11.3). There was less precipitation in 1988 and this was probably the cause of the lower fruit set. By standardizing the decrease in percentage between the two years it was verified that the infected shrubs produced more fruits than uninfected shrubs. The detrimental effect observed by infection by *T. laetus*, in terms of reproductive effort, will be compensated for by the effect produced as a result of infection by *E. colliguayae* (Figure 11.4). Both gall-formers affect plant architecture therefore, but in different ways.

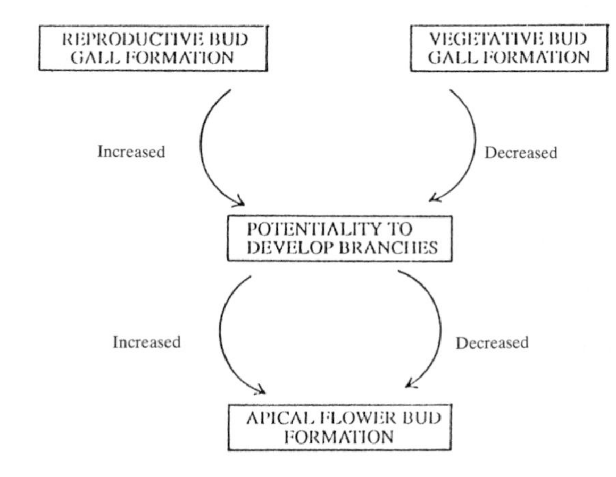

Fig. 11.4. Diagram of relationship between the effect of two gall-former insects and *Colliguaja odorifera* shrubs.

Concluding remarks

A knowledge of metameric architecture is an important tool for management. Plant species in Mediterranean-climate areas of central Chile, France, Israel and South Africa developed similar modular types of construction that allow them to optimize community resource use within the constraints of irradiance, temperature, annual precipitation and soil moisture (assuming, of course, that resource use is important for species survival).

The morphology and development over time of the modules formed showed similar trends in different growth forms, despite dissimilar phylogenetic histories.

Insects that form galls or eat buds affect plant architecture in a major way, although allocation of resources appears to be compensated for by the plant.

Acknowledgments

This work was supported by Grant FONDECYT 747/91 Mellon Foundation and CEE No. 918049 to G. Montenegro. We express our appreciation to Fundacion Andes for travel support to attend the VI International MEDECOS meeting in Crete, Greece. R. Ginocchio was supported by a Fundacion Andes doctoral fellowship.

References

Abrahamson, W.G. & Weis, A.E. 1978. Nutritional ecology of arthropod gall makers. In: Slansky, F. & Rodriguez, J.G. (eds.), Nutritional Ecology of Insects, Mites and Spiders. John Wiley and Sons Inc., New York, USA. pp. 236–258.

Aljaro, M.E., Frías, D. & Montenegro, G. 1984. Life cycle of *Rachiptera limbata* (Diptera: Tephritidea) and its relationship with *Baccharis linearis* (Compositae). Rev. Ch. Hist. Nat. 57: 123–129.

Chew, R.M. 1974. Consumers as regulators of ecosystems: an alternative to energetics. Ohio J. Sci. 74: 359–369.

Chew, R.M. & Chew, A.E. 1970. Energy relationships of mammals of a desert shrub. Ecol. Monogr. 40: 1–21.

Christensen, N.L. & Muller, C.H. 1975. Relative importance of factors controlling germination and seedling survival in *Adenostoma* chaparral. Am. Midl. Nat. 93: 71–78.

Etchegaray, J. & Fuentes, E.R. 1980. Insectos defoliadores asociados a siete especies arbustivas del matorral. An. Mus. Hist. Nat. 13: 159–166.

Everitt, B.S. 1977. The Analysis of Contingency Tables. Halsted Press, New York.

Flores, E., Mujica, A.M., Gómez, M. & Montenegro, G. 1987. Morfoanatomía de órganos vegetativos en *Schinus polygamus* y relación con el daño por insectos formadores de agallas. Boletín XXX Reunión Anual de la Sociedad de Biología de Chile, La Serena.

Floret, Ch., Galan, M.Y., Le Floc'h, L.E., Leprince, F. & Romane, F. 1989. France. In: Orshan, G. (ed.), Plant Phenomorphological Studies in Mediterranean Type Ecosystems. Kluwer Academic Publishers, Dordrecht, The Netherlands. pp. 9–98.

Franco, M., 1986. The influence of neighbours on the growth of modular organisms with an example from trees. In: Harper, J.L., F.R.S., Rosen, B.R. & White, J. (eds.), The Growth and Form of Modular Organisms. Phil. Trans. R. Soc. Lond. B313. pp. 209–225.

Frazer, J.M. & Davis, S.D. 1988. Differential survival of chaparral seedlings during the first summer drought after wild fire. Oecologia 76: 215–221.

Fuentes, E.R., Etchegaray, J., Aljaro, M.E. & Montenegro, G. 1981. Shrub defoliation by matorral insects. In: Di Castri, F., Goodall D.W. & Specht, R.L. (eds.), Mediterranean-Type Shrublands. Elsevier Scientific Publishing Co., Amsterdam, The Netherlands. pp. 345–359.

Fuentes, E.R., Poaini, A. & Molina, J.D. 1987. Shrub defoliation in the Chilean matorral: What is its significance? Rev. Ch. Hist. Nat. 60: 276–283.

Gashwiller, J.S. 1970. Further study of conifer seed survival in a western Oregon clearcut. Ecology 51: 849–854.

Ginocchio, R. & Montenegro, G. 1992. Interpretation of metameric architecture in dominant shrubs of the Chilean matorral. Oecologia 90: 451–456.

Ginocchio, R., Holmgren, M. & Montenegro, G. 1993. Effect of fire on plant architecture. Rev. Ch. Hist. Nat. (submitted).

Golley, F.B. 1973. Impact of small mammals on primary production. In: Gessman, J.A. (ed.), Ecological Energetics of Homeotherms. Utah State Univ. Monograph Series V.20. pp. 142–147.

Grez, A. 1988. *Procalus malaiseii* y *Procalus lenzi* (Coleoptera: Chrysomelidae): dos especialistas dell matorral. Rev. Ch. Entomología 16: 65–67.

Hallé, F. 1978. Architectural variations at the specific level in tropical trees. In: Tomlinson, P.B. & Zimmermann, M.H. (eds.), Tropical Trees as Living Systems, Cambridge University Press, United Kingdom. pp. 209–221.

Hallé, F. 1986. Modular growth in seed plants. In: Harper, J.L., F.R.S., Rosen, B.R. & White, J. (eds.), The Growth and Form of Modular Organisms. Phil. Trans. R. Soc. Lond. B313. pp. 77–87.

Hallé, F. & Oldeman, R.A.A. 1970. Essai sur l'architecture et la dynamique de croissance des arbres tropicaux, Masson, Paris.

Hallé, F., Oldeman, R.A.A. & Tomlinson, P.B. (eds.). 1978. Tropical Trees and Forests: an Architectural Analysis. Springer-Verlag, New York, USA.

Hardwick, R.C. 1986. Physiological consequences of modular growth in plants. In: Harper, J.L., F.R.S., Rosen, B.R. & White, J. (eds.), The Growth and Form of Modular Organisms. Phil. Trans. R. Soc. Lond. B313. pp. 161–173.

Harper, J.L. 1977. Population Biology of Plants. Academic Press, London, New York, San Francisco.

Harper, J.L. & Bell, A.D. 1979. The population dynamics of growth form in organisms with modular construction. In: Anderson, R.M., Turner, B.D. & Taylor, L.R. (eds.), Population Dynamics. 20th Symposium of the British Ecological Sociey. Blackwell Scientific Publications, Oxford. pp. 29–52.

Hilbert, D.W., Swift, D.M., Detling, J.K. & Dyer, M.I. 1981. Relative growth rates and the grazing optimization hypothesis. Oecologia 51: 14–18.

Hoffmann, A.J. 1972. Morphology and histology of *Trevoa trinervis* (Rhamnaceae) a drought-deciduous shrub from the Chilean matorral. Flora 16: 527–538.

Hoffmann, A.J. & Hoffmann, A.E. 1976. Growth pattern and seasonal behavior of buds of *Colliguaja odorifera*, a shrub from the Chilean mediterranean vegetation. Can. J. Bot. 1767–1774.

Janzen, D.H. 1969. Seed-eaters versus seed size, number, toxicity and dispersal. Evolution 23: 1–27.

Küppers, M. 1989. Ecological significance of above ground architectural patterns in woody plants: a question of cost-benefit relationship. Tree 4: 375–379.

Le Roux, A., Perry, P. & Kyriacou X. 1989. South Africa. In: Plant Phenomorphological Studies in Mediterranean Type Ecosystems. Kluwer Academic Publishers, Dordrecht, The Netherlands. pp. 159–346.

Louda, S.M. 1983. Seed predation and seedling mortality in the recruitment of a shrub, *Haploppapus venetus* (Asteraceae), along a climatic gradient. Ecology 64: 511–521.

Lowell, P.H. & Lowell, P.J. 1985. The importance of plant form as a determining factor in competition and habitat exploitation. In: White, J. (ed.), Studies on Plant Demography: a festschrift for John L. Harper. Academic Press, London. pp. 209–221.

Martínez, E., Montenegro, G. & Elgueta, M. 1992. Distribution and abundance of two gall makers on the Euphorbiaceous shrub *Colliguaja odorifera*. Rev. Ch. Hist. Nat. 65: 75–82.

McNaughton, S.J. 1976. Serengeti migratory wildebeest: facilitation of energy flow by grazing. Science 191: 92–94.

McNaughton, S.J. 1979. Grazing as an optimization process: grass-ungulate relationships in the Serengeti. Am. Nat. 113: 691–703.

McNaughton, S.J. 1983. Compensatory plant growth as a response to herbivory. Oikos 40: 329–336.

McNaughton, S.J. 1984. Grazing lawns: animals in herds, plant form and coevolution. Am. Nat. 124: 863–886.

Mills, J.N. 1983. Herbivory and seedling establishment in post-fire southern California chaparral. Oecologia 60: 267–270.

Montenegro, G., Aljaro, M.E. & Kummerow, J. 1979. Growth dynamics of Chilean matorral shrubs. Bot. Gaz. 140: 114–119.

Montenegro, G., Jordán, M. & Aljaro, M.E. 1980. Interactions between Chilean matorral shrubs and phytophagous insects. Oecologia (Berlin) 45: 346–349.

Montenegro, G., Aljaro, M.E., Avila, G. & Mujica, A.M. 1988. Growth patterns as determined by water stress and adaptation. In: Di Castri, F., Floret, Ch., Rambal, S. & Roy, J. (eds.), Time Scales and Water Stress. Proc. 5th International Conference on Mediterranean Ecosystems. I.U.B.S., Paris, France. pp. 277–285.

Montenegro, G., Avila, G., Aljaro, M.E., Osorio, R. & Gómez, M. 1989. Chile. In: Orshan, G. (ed.), Plant Phenomorphological Studies in Mediterranean Type Ecosystems. Kluwer Academic Publishers, Dordrecht, The Netherlands. pp. 347–388.

Morrow, P.A. 1983. The role of sclerophyllous leaves in determining insect grazing damage. In: Kruger, F.J., Mitchell, D.T. & Jarvis, J.U.M. (eds.), Mediterranean-Type Ecosystems: The Role of Nutrients. Springer Verlag, Berlin. pp. 509–524.

Morrow, P.A. & La Marche, V.C. 1978. Tree ring evidence for chronic insect suppression of productivity in subalpine *Eucalyptus*. Science 201: 1244–1246.

Oechel, W.C., Lawrence, W., Mustafa, J. & Martínez J. 1981. Energy and carbon acquisition. In: Miller, P.C. (ed.), Resource Use by Chaparral and Matorral. A Comparison of Vegetation Function in Two Mediterranean Type Ecosystems. Springer-Verlag, New York. pp.151–183.

Orians, G.H. & Solbrig, O.T. 1977. A cost-income model of leaves and roots with spatial reference to arid and semiarid areas. Am. Nat. 111: 677–690.

Orshan, G. 1989a. Israel. In: Orshan, G. (ed.), Plant Phenomorphological Studies in Mediterranean Type Ecosystems. Kluwer Academic Publishers, Dordrecht, The Netherlands. pp. 99–158.

Orshan, G. (ed.), 1989b. Plant Phenomorphological Studies in Mediterranean Type Ecosystems. Kluwer Academic Publishers, Dordrecht, The Netherlands.

Owen, D.F. 1980. How do plants benefit from the animals that eat them? Oikos 35: 230–235.

Owen, D.F. & Weigert, R.G. 1976. Do consumers maximize plant fitness? Oikos 27: 488–492.

Prevost, M.F. 1966. Architecture de quelques apocynacées ligneuses. Bull. Soc. Bot. Fr. 114: 23–36.

Putwain, P.D. & Harper, J.L. 1972. Studies in the dynamics of plant populations V. Mechanisms governing the sex ratios in *Rumex acetosa* and *R. acetosella*. J. Ecol. 60: 113–129.

Quinn, R. 1986. Mammalian herbivory and resilience in mediterranean-type ecosystems. In: Dell, B., Hopkins, A.J. & Lamont, B.B. (eds.), Resilience in Mediterranean Ecosystems. Dr W Junk Publ., Dordrecht, The Netherlands. pp. 113–128.

Solervicens, J. & Elgueta, M. 1989. Entomofauna asociada al matorral costero del Norte Chico. Acta Ent. Chilena 15: 19–122.

Walkowiak, A.M., Simonetti, J.A., Serey, I., Jordán, M., Arranz, R. & Montenegro, G. 1984. Defensive patterns in shrubs of central Chile; a common strategy? Acta Oecol. Plant. 15: 191–199.

Watkinson, A.R. & White, J. 1986. Some life-history consequences of modular construction in plants. In: Harper, J.L., F.R.S., Rosen, B.R. & White, J. (eds.), The Growth and Form of Modular Organisms. Phil. Trans. R. Soc. London B313. pp. 31–51.

White, J. 1979. The plant as a metapopulation. Ann. Rev. Ecol. Syst. 10: 109–145.

White, J. 1984. Plant metamerism. In: Dirzo, R. & Sarukhán, J. (eds.), Perspectives on Plant Population Ecology. Sinauer Associates, Sunderlands, Massachusetts. pp. 15–47.

PART FIVE

Pollination

CHAPTER 12

Pollination syndromes in the Mediterranean: generalizations and peculiarities

AMOTS DAFNI[1] and CHRISTOPHER O'TOOLE[2]
[1]Institute of Evolution, Haifa University, Haifa 31905, Israel
[2]Hope Entomological Collections, University Museum, Oxford OX1 3PW, UK

Key words: core plants, flower colour, maquis, Mediterranean, regional peculiarities, pollination biology, pollination market, seasonal features

Abstract. We review the general and peculiar features of pollination syndromes in the Mediterranean, with special emphasis on the maquis, and the role of flower colour and core plants at the community level. We emphasize the need for multinational, coordinated research on the pollination biology of the Mediterranean region and the associated need for a greater understanding of and investment in fundamental research on the taxonomy of bees.

Introduction

Characteristics of the Mediterranean region and its flora

The limits of the Mediterranean region have been studied from various standpoints: floristic, vegetational, climatic and bioclimatic. It is not surprising, therefore, that various authors define the limits of the Mediterranean region in different ways. Nevertheless, there is broad agreement that a climate may be considered to be Mediterranean if it has the following characteristics: 1) the summer is the driest season and 2) there is a period of effective physiological drought (Quezel, 1985).

The Mediterranean flora comprises about 25,000 species in an area of 2,300,000 × 10^3 km^2 (Quezel, 1985). The first autumn rains break the long summer drought, first in the upper soil layers. At this time, the annuals and shallow-rooted perennial herbs and subshrubs begin their seasonal growth cycle. The growth cycle of the deeper-rooted evergreen shrubs does not generally begin until spring (Herrera, 1986).

The peak of flowering is in the spring (March–April) when most of the annuals bloom, whilst deep-rooted perennials and geophytes may extend their flowering season also into the summer and the autumn (Kummerow, 1983).

Many typical Mediterranean shrubs (e.g. *Cytisus*, *Cistus*, *Lavandula*, *Thymus*), flower ing the transition from the wet to the dry son (April–May). All these plants have shallow roots, lack storage organs and tend to occupy exposed habitats (Herrera, 1986). Herrera suggests that these characteristics force the shrubs into a 'spring quick-ripening phenological strategy'. This consideration may also apply to the annuals which also peak in this season and are very typical of the Mediterranean (Shmida, 1981).

Human activities, especially during the last millennium caused a serious degradation of the Mediterranean maquis (Pignatti, 1978). On the other hand, road construction, the establishment of paths, tracks, unpaved roads, sand and gravel pits and woodland edge has maintained and even increased the range of nest sites for wild bee populations. In other words, disturbed habitats are now centres of richness and diversity for wild bees (Fig. 12.1).

M. Arianoutsou and R.H. Groves, Plant-Animal Interactions in Mediterranean-Type Ecosystems, 125–135, 1994.

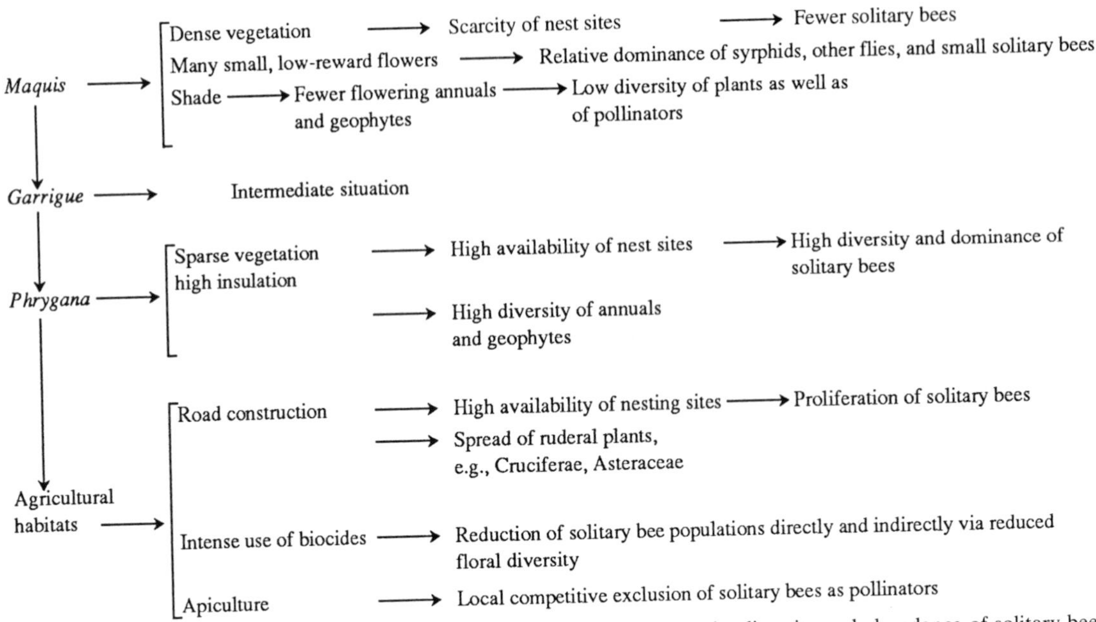

Fig. 12.1. Degradation of the Mediterranean ecosystems and its likely effects on the diversity and abundance of solitary bees.

Intensive land use and the increasing use of herbicides and insecticides have caused an apparent reduction of wild bee populations in northern latitudes (Grebbenikov, 1972; Williams, 1989; Corbet *et al.*, 1991). Although there is no direct evidence for the Mediterranean, it is reasonable to assume that a similar decline in wild bee populations can be expected in areas where agriculture is becoming more intensive, where insecticide usage is increasing and where there is coastal development associated with the tourist industry.

In areas where apiculture is practised intensively, there may well be adverse competitive effects on wild bee populations. Although there are no reliable data for the Mediterranean area, Roubik (1988) has documented and discussed competitive interactions between introduced *Apis mellifera*, Africanized and otherwise, with native bees in the neotropics. O'Toole (unpubl.) found that in Israel, the numbers of wild bees, mainly species of *Andrena*, *Lasioglossum*, *Halictus* and *Eucera* foraging at ruderal Cruciferae were markedly lower in areas close to kibbutzes where apiculture is practised.

The Mediterranean Basin is a noted centre of diversity of bee species (Michener, 1979; O'Toole, 1991; O'Toole & Raw, 1991). The final inventory of wild bees for this region may comprise 3000–4000 species (Baker, pers. comm.).

Bee genera which are particularly speciose in the Mediterranean are *Andrena* (Andrenidae), *Lasioglossum* and *Halictus* (Halictidae), *Anthophora* and *Eucera* (Anthophoridae), and *Hoplitis*, *Anthocopa*, *Megachile* and *Chalicodoma* (Megachilidae).

In general, the Mediterranean bee fauna can be divided into three groups: circum-Mediterranean, West Mediterranean and East Mediterranean. In Israel, a proportion of the bees occurring in the Mediterranean zone are of Caucasian or trans-Caspian origin. Generalizations of this sort should, however, be seen in the context of the relatively poor state of taxonomic knowledge of the Mediterranean bee fauna.

The Mediterranean flora as a whole is dominated by species which are pollinated largely by wild bees: Spain (Herrera, 1987), Greece (Petanidou & Vokou, 1990), Israel (Dukas & Shmida, 1990).

In relation to their foraging behaviour, the Mediterranean bees can be divided into two broad categories:

1) Small bees (3–7 mm long) with short tongues (0.5–4 mm) which are slow-flying, short-distance foragers, with relatively low energetic demands, associated with open access, low reward flowers.
2) Medium to large bees (10–25 mm long) with long tongues (8–23 mm long) which are fast-flying, long-distance foragers, with high energetic demands, associated with restricted access, high reward flowers (O'Toole & Shmida, unpubl.).

Other insects such as bee-flies (Bombyliidae), hoverflies (Syrphidae), wasps, butterflies and hawkmoths (Sphingidae) are relatively unimportant as pollinators. Flies are relatively more frequent flower visitors at the end of the summer but wild bees still dominate as effective pollinators.

The pollination market throughout the year

The factors which determine the nature of the 'pollination market' (Selten & Shmida, 1991) change through the year, thereby imposing different pressures on advertisement (flower size) and reward in the various seasons.

The spring peak of flowering produces a surplus of flowers relative to the pollinators, resulting in high competition between flowers for pollinators. The result is a large investment in reward (nectar and pollen) and advertisement (large, colourful flowers) (Cohen & Shmida, 1993).

In general, the spring pollination market is a 'buyers' market', and is driven by the needs of the pollinators (Table 12.1). Thus, it is expected that the plants will invest more (relative to vegetative structures) in their flowers (advertisement and reward), leading to 'competitive advertisement'. This may explain why spring flowers are generally larger and more rewarding than flowers of any other season (Cohen & Shmida, 1993).

At the onset of summer (May–June) the situation in the Mediterranean 'pollination market' is sharply reversed. A shift occurs from a 'buyers' market' to a 'sellers' market'. That is, there is an evident surplus of insects over flowers. Under these conditions, one would expect to find a reduction in floral investment reward as well as advertisement (Shmida & Dafni, 1989).

Such a trend is indeed found and is well-exemplified by the Lamiaceae in Israel, which show a progressive seasonal decrease in flower size from the spring onwards (Dukas & Shmida, 1990). This is accompanied by a parallel reduction of the flower advertising area and reward (Dafni, 1990).

In the autumn (October–December), there is a 'truncated pollination market', characterized by relatively few flowering species pollinated by a reduced number of pollinators, mainly small bees. Typical of this season are the showy flowers of large, stemless geophytes (*Colchicum* spp. *Crocus* spp.), which appear in low abundance.

The appearance of large flowers is explained in terms of 'discovery advertisement' (Shmida & Dafni, 1989). That is, at a time when the few species in flower are at low densities, the most important factor in attracting pollinators over a long distance is a large visual signal and this contrasts with the reward-based 'competitive advertisement' of spring and early summer. Thus, the large flowers engaged in 'discovery advertisement' can be expected to have, and, indeed, do have a relatively low reward.

By contrast, 'competitive', reward-based advertisement predominates in spring, when a market of many flowering species competes for the pollination services of a rich and diverse bee fauna.

Some autumn flowering geophytes have 'combined pollination mechanisms' by offering simultaneously pollen to hoverflies (Syrphidae) and pollen and nectar to small bees, e.g., *Sternbergia clusiana* (Dafni & Werker, 1982), *Pancratium parviflorum* and *Colchicum* spp. (Dafni, unpubl.). This is interpreted as an adaptation to maximise pollination in a harsh season with few pollinators (Herrera, 1982; Dafni & Werker, 1982).

Although this general model reflects the situation which prevails in Israel, other authors also indicate a peak of insect activity in spring in other parts of the Mediterranean and their scarcity at the end of summer (Spain: C. Herrera, 1982, J. Herrera, 1986; Greece: Petanidou, pers. comm.), which enables us to extend our considerations to embrace the whole Mediterranean region.

Table 12.1. The relative importance of various pollinators throughout the year in the Mediterranean 'pollination market' in Israel

Pollinator group	Spring	Summer	Autumn	Winter
Butterflies	+	−	±	±
Sphingidae (hawkmoths)	±	±	−	−
Diptera (flies)	+	+ +	+ +	±
Bombyliidae (bee–flies)	+	−	−	−
Coleoptera (beetles)	+	−	−	−
Vespidae (social wasps)	±	±	−	−
Solitary bees (small)	+ +	+ + +	+	+
Solitary bees (large)	+ +	+	±	±

Key as Table 12.4.

The Mediterranean maquis as a pollination environment

An analysis of the flower characteristics, flowering season, flower sexuality and pollination mechanism of the main components of the Mediterranean maquis by Dafni (1986) shows (Tables 12.2 and 12.3) the following:

1) Small, inconspicuous flowers: 35% of the species are small (<10 mm in diameter) and inconspicuous (green, yellow and cream-coloured flowers).
2) Floral structure (exposed nectar and lack of mechanisms to limit pollinator diversity) suggests pollination by flies, including hoverflies (Syrphidae) and small bees.
3) Forty per cent of the small flowers are unisexual, some of which are dioecious, a situation which imposes limitations on the pollination mechanisms and reward partitioning between the sexes. Generally, the male plant offers nectar as well as pollen and the female offers only nectar, sometimes nothing at all, which can reduce the frequency of visits to the female plant. J. Herrera (1986) suggests that presenting many small unspecialized, low reward flowers may be a means by which scrub plants maximise fruit set.
4) White flowers: white flowers are prominent against a dark background such as forest floor (Baker & Hurd, 1968). An analysis of flower colour in the maquis shows the following: small green flowers (16 examples), white flowers larger than 1 cm in diameter (11), colourful (5), wind-pollinated (11) and others (4). This emphasizes the rarity of colourful flowers in the maquis in comparison with the phrygana. While some herbaceous species in the maquis have white or whitish flowers (e.g. *Galium samuelssonii*, *Silene italica*, *S. vulgaris*, *S. trinervis*, *Allium subhirsutum* and *A. neapolitanicum*), their congeners in other habitats are colourful.
5) Mass, gregarious flowering: this is common and typical of tropical forests (Bawa, 1983). In the Mediterranean maquis, conspicuous massive flowering occurs in *Cercis siliquastrum*, *Crataegus* spp., *Prunus ursinum*, *Pyrus syriaca*, *Eriolobus trilobatus* and *Amygdalus communis*. In all these species, the individuals are widely dispersed and the high seed set indicates the success of this attractive mass flowering strategy.
6) Wind pollination. In Israel, the dominant species of the maquis (*Quercus* spp., *Pistacia* spp.) are wind-pollinated. This is also true of the maquis of Greece (Petanidou & Vokou, 1990). Typically, flowering precedes leafing in the evergreen species, but not so prominently as in the deciduous species. Twelve species which are components of the maquis are wind-pollinated but at least three (*Ceratonia siliqua*, *Olea europaea* and *Rhamnus alaternus*) are also insect-pollinated.

The dominant assemblage of pollination mechanisms in the Mediterranean maquis can be regarded as a snapshot of pollination syndromes selected in various biogeographical regions. The fact that they co-exist in the same habitat suggests that there are no serious limitations on pollination.

There is no current evidence of any special 'maquis pollination syndrome'. Many of the species are regarded as 'generalists', which implies that pollination is not a crucial limiting factor in the maquis environment. J. Herrera (1986)

Table 12.2. Distribution, flower characteristics, flower season, leaf duration and pollination mode in maquis trees and shrubs

Species	Distribution	Life form	Flower size (cm)	Flower colour	Nectar	Scent	Flower sexuality	Unisexual flower distribution	Flowering season (months)	Leaf duration	Mode of pollination
Acer obtusifolium Sm.	EM	T	0.5–0.7	G–Y	+	?	U,B	M	4–5	E	I
Arbutus andrachne L.	EM	T	1.1–1.3	W	+	+	B	–	3–4	E	I
Celtis australis L.	M	T	0.6–0.9	G–Y	?	–	A	–	3–4	D	I
Ceratonia siliqua L.	M	T	1.1–1.3	B–R	+	+	U,D	D	9–11	E	I(W)
Cercis siliquastrum L.	EM–WM	T	2.0–3.0	P	+	–	B	–	3–4	D	I
Crataegus aronia (L.) D.C.	EM–WIT	T	0.5–0.7	W	+	+	B	–	3–4	D	I
Cragaegus azarolus L.	EM	T	1.3–1.5	W	+	+	B	–	4–5	D	I
Eriolobus trilobatus (Poiret) M. Roemer	EM	T	3.0–5.0	W	+	+	B	–	5	D	I
Juniperus oxycedrus L.	M	T	–	–	–	–	US	D	3–4	E	W
Laurus nobilis L.	M	T	1.1–1.2	W–Y	+	+	U	D	3–6	E	I
Myrtus communis L.	M	SH	1.3–1.5	W	–	+	B	–	4–8	E	I
Olea europaea L.	M	Y	0.5–0.6	W–Y	–	+	B	–	4–5	E	I(W)
Phillyrea latifolia L.	M	SH	0.5–0.6	W–G	–	+	B	–	3–4	E	I(W)
Pistacia lentiscus L.	M	SH	0.5–0.7	G–W	–	–	U	D	3–4	E	W
P. palaestina Boiss.	EM	T	0.5–0.7	G–R	–	–	U	D	3–4	D	W
Prunus ursina Ky.	EM	T	0.6–1.0	W	+	+	B	–	3–4	D	I
Pyrus syriaca Boiss.	EM–WIT	T	2.0–2.5	W	+	+	B	–	3–4	D	I
Quercus calliprinos Webb	EM	T	0.3–0.5	G	–	–	U	M	2–3	E	W
Q. boissierii Reut.	EM	T	0.3–0.5	G	–	–	U	M	3–4	D	W
Q. ithaburensis Decne.	EM	T	0.3–0.5	G	–	–	U	M	3–4	D	W
Rhamnus alaternus L.	EM	T	0.3–0.4	Y–G	+	+	U	D	2–4	D	I(W)
Rh. lycoides L. spp. *graeca* (Boiss et Reuter) Tutin	EM	SH	0.2–0.3	Y–G	+	+	U	D	3–4	D	I
Rh. punctata Boiss.	M	T	0.2–0.3	Y–G	+	+	U	D	3–4	D	I
Rhus coriaria L.	IT	SH	0.5–0.7	G–Y	?	–	U	M	5–6	D	I
Rosa canina L.	ES–MIT	SH	2.0–2.5	P	+	+	B	–	5–8	D	I
Styrax officinalis L.	EM	T	2.0–3.0	W	–	–	B	–	4–6	D	I
Viburnum tinus L.	M	SH	0.6–0.9	W	+	+	B	–	3–4	E	I
Ulmus minor Miller ssp. *canescens* (Melv.) Brow. et Ziel.	NM–EM	SH–T	0.5–0.7	G–Y	?	+	U,B	M	2–3	D	I

Distribution

EM = East Mediterranean
M = Circum-Mediterranean
NM = North Mediterranean
IT = Irano-Turanian
ES = Euro-Siberian

Flower colour

G = Green
Y = Yellow
B = Brown
R = Red
W = White
P = Pink

Flower sexuality

U = Unisexual
US = Unisexual strobili
B = Bisexual

Mode of pollination

W = Wind pollination
I = Insect pollination

Unisexual flower

M = Monoecious
D = Dioecious
A = Andromonoecious

Leaf duration distribution

E = Evergreen
B = Deciduous
SD = Summer deciduous

Life form

T = Tree
SH = Shrub
C = Climber
G = Geophyte

Table 12.3. Distribution, life form, flower characteristics, flowering season, leaf duration and pollination mode in Mediterranean climbers

Species	Distribution	Life form	Flower size (cm)	Flower colour	Nectar	Scent	Flower sexuality	Unisexual flower distribution	Flowering season (months)	Leaf duration	Mode of pollination
Aristolochia sempervirens L.	M	C	2.5–3.0	Y–V	+	+	B	–	4–6	SD	I
A. billardieri Jaub. et Spach	EM	C	3.0–4.0	P–Y	+	+	B	–	3–5	E	I
Asparagus aphyllus L.	M	C	0.2–0.3	G–W	+	–	B	–	9–11	E	I
Bryonia cretica L.	EM	C	0.5–0.7	Y–G	?	–	U	D	3–5	E	I
B. syriaca Boiss.	EM	C	1.0–1.5	Y–G	+	+	U	D,M	3–5	E	I
Clematis cirrhosa L.	M	C	4.0–6.0	Y–G	+	+	B	–	10–2	SD	I
C. flammula L.	M–IT	C	2.0–2.5	W	+	–	B	–	4–7	E	I
Ephedra foemina Forrsk.	EM	C	0.5–0.75	G–W	+	–	U	D	4–10	SD	W
Hedra helix L.	ES–M	C	0.3–0.4	G–W	+	+	B	–	9–11	E	I
Lonicera etrusca Santi	M	C	3.0–4.0	W–Y	+	+	B	–	4–5	D	I
Prasium majus L.	M	SH	1.0–1.5	W	+	–	B	–	1–7	E	I
Rubus canescens D.C.	NM	SH	0.3–0.9	W	?	–	B	–	4–6	D	I
Smilax aspera L.	M	C	0.3–0.4	Y–G	+	–	B	–	10–11	E	I
Tamus communis L.	M	C,G	0.3–0.4	W–P	+	+	U	D	2–4	D	I
T. orientalis Thieb.	EM	C,G	0.3–0.4	WP	+	+	U	D	12–1	D	I

See for abbreviations Table 12.2.

Table 12.4. The relative importance of various pollinator groups as pollinators in Mediterranean vegetation types.

Pollinator type	Maquis	Garrigue	Phrygana
Butterflies	–	+	+ +
Sphingidae (hawkmoths)	±	–	–
Bombyliidae (bee–flies)	+	+	+
Syrphidae (hoverflies)	+ + +	+ +	+ +
Coleoptera (beetles)	–	+	+ +
Vespidae (social wasps)	+ +	+	±
Solitary bees (small)	+ + +	+ +	+
Solitary bees (large)	±	+ +	+ + +

Key:
– = unimportant
± = incidental
+ = common
+ + = important
+ + + = dominant

recognizes the same trend and explains that 'fruit and seed features may have been more significant in the evolutionary history of the Mediterranean scrub plants than pollination mechanisms'.

General surveys of the pollinators of the maquis in Israel (Dafni, O'Toole, Kasher, Shmida, unpubl.) show that hoverflies (Syrphidae), wasps and small solitary and primitively social bees are the main pollinators (Table 12.4). J. Herrera (1985, 1987) has shown that in many instances, the scrub species offer mainly pollen and only small amounts of nectar, if any at all. Most of the pollinators are flies, small, non-specialized beetles, and short-tongued bees, all with low energetic demands.

Flower colour

Dafni and Shmida (unpubl.) surveyed the range of colours found in the flora of Israel (c. 2000 spp.). The four dominant colours were pink-purple (21.1%), yellow (20%), white (15.1%) and cream (7.0%). They obtained similar results when the analysis was made for dominant species or on a month-by-month basis.

Menzel and Shmida (1993) report a detailed analysis of the flower reflectance spectrum for the flora of Israel, including the ultraviolet (UV) range, for 350 species in relation to their form, symmetry, quality of reward and pollinators. They found two main syndromes of floral traits which show correlation with pollinators:

1) Sharp, distinct UV-contrast but without fine patterning; these flowers are pollinated by insects with low energy requirements, mainly flies and small bees.
2) Fine, symmetrical UV contrast patterns, the UV dark or with gradual pale-dark bicoloration; these flowers are pollinated by large bees with high energy requirements, which can detect and process fine visual cues.

Dukas and Shmida (1990) examined the relations between the colour, size and shape of Israeli cruciferous flowers and their pollinators. They found that violet flowers were larger, tubular and pollinated by large, long-tongued bees. Yellow and white flowers are smaller, with open access (i.e., non-tubular) and are pollinated by flies and short-tongued, small bees.

J. Herrera (1987) noted that many yellow or white-flowered taxa offer pollen as the main reward, while nectar is the main reward in pink flowers. This generalization seems to hold for the Israeli flora as a whole and agrees with the findings of Frankie *et al.* (1983) from Costa Rica. They showed that large bees visit mainly large, violet or pink, bilaterally symmetrical flowers, whilst smaller bees tend to visit yellow or white, radially symmetrical flowers.

Because solitary and primitively social bees are the dominant pollinators of native floras in the Mediterranean Basin, it is likely that the flower colour spectrum and patterns of UV reflectance are the result of co-evolution with this group of pollinators.

Although there is broad agreement with the world colour spectrum (Weevers, 1952) a more detailed, local analysis of floral colour spectra, including UV reflectance patterns, is clearly necessary before drawing any further conclusions.

'Core plants'

Several mass flowering species occur in open areas of *phrygana* and *garrigue*. These species, e.g., *Salvia fruticosa*, *Asphodelus aestivus*, *Lavandula stoechas*, and *Coridothymus capitatus* flower in large populations, produce partially concealed but ample exposed pollen. In Israel, the flowers of these species are visited by many species of solitary bees in the genera *Anthophora*, *Habropoda*, *Eucera*, *Synhalonia*, *Cubitalia* and *Chalicodoma* and by primitively social bees in the genus *Halictus*. Only a few of these bees are real pollinators. Thus, there is a high ratio of visitors to pollinators; these plants attract and maintain large populations of solitary bees which pollinate other plant species. We base this conclusion on the distribution of unconsolidated pollen on the bodies of many species of bees visiting the above mentioned species, in relation to the chances of the pollen brushing against the target stigmas of these plants (Dafni, unpubl.; Fig. 12.2).

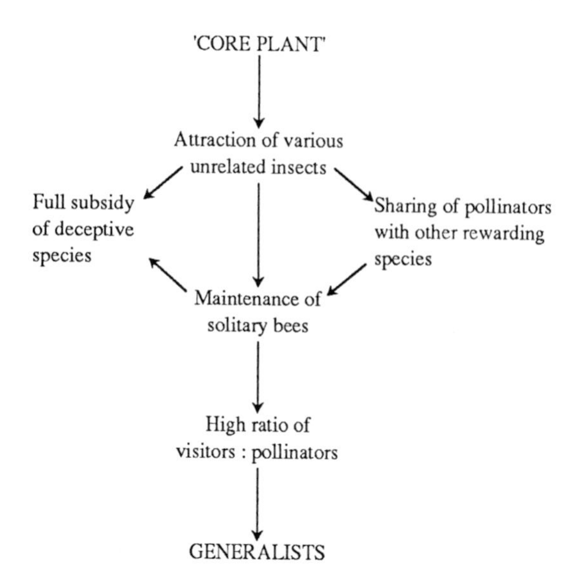

Fig. 12.2. "Core plants" in the Mediterranean phrygana and maquis and their role in pollination at the community level.

Some flowering plants share pollinators, with the result that while a given bee species might gather nectar from a 'core plant', it uses other plants as pollen sources. This is particularly apparent with the many bee species visiting Lamiaceae and Boraginaceae. Here, the flowers have concealed nectar and, with the exception of specialist pollinators, pollen deposition on the bees is passive or incidental. In such cases, there is a mutual benefit for all the plant species involved, although 'core plants' are responsible for the main advertisement and attraction to the area.

Nectariferous 'core plants' fully subsidize nectarless plants which rely on deception for pollination purposes. The deceptive species are general mimics of a nectariferous plant (Dafni, 1984) and the model itself is a generalist. As a result, several species of 'core plants' may be involved in the pollination of one *Orchis* species, e.g., *Orchis caspia* 'borrows' its pollinators from *Asphodelus aestivus* as well as from *Salvia fruticosa* (Dafni, 1983). Most of the deceptive species are in flower at the same time as at least one 'core' species.

Vogel and Westerkamp (1991) refer to the same phenomenon and call these plants 'magnet species', i.e., they are similar but commoner species that attract pollinators and share them with the rarer species at the expense of mixed pollination.

Peculiarities of the Mediterranean

Deception

Deceptive flowers are represented mainly by the Orchidaceae [but see J. Herrera (1991) for *Nerium oleander*]. Food deceptive syndromes are common in the genera *Orchis*, *Dactylorrhiza* and *Cephalanthera* and exemplify such types as Batesian mimicry, guild-mimic deception and non-mimetic deception (Dafni, 1984).

Reproductive deception includes sexual deceit, especially the well-known induction of pseudocopulation among male wasps and bees by species of *Ophrys* (see Dafni, 1987; Dafni & Bernhardt, 1990; Paulus & Gack, 1990, for reviews of the subject).

Why the Mediterranean should be so rich in deceptive species remains an open question. The fact that most of the pollinators of these species are solitary bees which are connected with dry, well-lit, open habitats, may suggest that it is a result of recent human intervention as was proposed for *Ophrys* by Kullenberg and Bergström (1976). Other explanations such as climatic fluctuations (Kullenberg & Bergström, 1976) or earliness of flowering as an adaptation to avoid long dry summers (Stebbins, 1970), are not supported by any evidence.

Red, bowl-shaped flowers

Red, bowl-shaped flowers in Israel form a peculiar guild of species pollinated by beetles of the genus *Amphicoma* (Glaphyridae) (Dafni *et al.*, 1990).

All of these plants (*Anemone coronaria*, *Ranunculus asiaticus*, *Tulipa agenensis*, *Papaver* spp. and *Adonis allepica*) have large (2–5 cm diameter), radially symmetrical, bowl-shaped flowers which are a conspicuous orange-red colour, with a black centre and massed, filamentous stamens. To the human antenna, these flowers are weakly-scented; they produce no nectar, their only reward being copious quantities of pollen. Visiting beetles become heavily dusted with pollen and it is clear that the pollination mechanism is of the simple 'mess and soil' type.

The members of this red-flowered 'poppy' guild flower in a sequence which is correlated with the amount of pollen produced by each of them and with the relative attractiveness to their pollinators.

Field experiments show that *Amphicoma* spp. beetles prefer red, odourless flower models to odourless models of other colours, and they prefer models with a dark centre to plain red ones (Dafni *et al.*, 1990). Later field experiments showed that the *Amphicoma* are attracted to red models based on the size of the model, regardless of its shape or depth. The increasing attraction of supra-normal sized models reflects the situation in the field, where members of the red-flower guild exhibit large, mass-flowering populations which present a long distance attraction signal to the pollinating beetles (Dafni, 1991).

Bees associated with 'non-bee flowers'

Bees are normally associated with sweet-scented, showy flowers. However, in the maquis of Israel, some bees seem to have shifted to plants usually pollinated by wasps and flies. For example, the andrenids *Andrena carmela* and *Melittoides melittoides* are associated with the figwort *Scrophularia rubricaulis* (O'Toole, unpubl.). Species of *Scrophularia* have small, inconspicuous green flowers with a reddish tinge and a foetid smell. They are normally regarded as adapted for pollination by social wasps, *Polistes* and *Vespula* spp. (Faegri & Van der Pijl, 1979). The females of *Andrena carmela* and *Melittoides melittoides* collect pollen from *Scrophularia rubricaulis*, the former species having been found only on this flower. The shift is clearly on the part of the bees: *S. rubricaulis* is visited by the expected social wasps.

The shift by some bees to flower types not normally attractive to bees seem to be a real phenomenon in the Mediterranean zone in Israel and requires further study.

Conclusions

The Mediterranean Basin has a rich and diverse flora of ca. 25,000 species and the dominant pollinators are bees, with an estimated 3000–4000 species.

The typical, eu-Mediterranean maquis elements, the evergreen sclerophyllous species, are derived from an old, tropical stock (Quezel, 1985) and have no special adaptations for pollination which could be regarded as unique to the Mediterranean.

The three pollination syndromes peculiar to the Mediterranean, large autumn flowers, deception, and red bowl-shaped flowers are, with the exception of *Papaver* spp., confined to geophytes, in which the Mediterranean is very rich.

The degradation of the dense maquis by humans encouraged the invasion of flowering plants which tolerate high insolation, many of which are annuals and geophytes. The same phenomena no doubt enhanced the spread and increased diversity of solitary bees in the degraded areas of maquis.

Epilogue: tasks for the future

This preliminary survey deals largely with those two extremes of the Mediterranean, Israel and Spain; much of the 'typical' Mediterranean region is *terra incognita* from the viewpoint of pollination biology. We therefore urge that a multinational research effort be established to co-ordinate studies at various foci within the region.

We take as our inspiration the Programa Cooperativo sobre la Apifauna Mexicana (PCAM). This United States–Mexican project has the following ends in view: to investigate bee behaviour and pollination biology, to prepare systematic revisions and keys to the bees, to establish a permanent reference collection of bees in Mexico and to establish a training programme for Mexican students (LaBerge, 1991).

Although this is primarily an entomology-led effort, it stands as a model for successful international collaboration and the basic philosophy behind it is relevant to the considerations we outline below. We suggest the following as suitable topics for a multinational effort:

1) Comparative pollination ecology of the same or closely related taxa throughout the region.
2) Convergence vs. divergence in pollination mechanisms of typical Mediterranean genera such as *Asphodelus*, *Narcissus*, *Silene*, *Pisum*, *Calicotome*, *Spartium*, *Genista*, *Cistus*, *Cyclamen*, *Teucrium*, *Salvia*, *Lavandula*, *Thymus* and *Linaria*.
3) A biological inventory of the bee fauna of the Mediterranean Basin, with special reference to flora relations. Central to this is the need to increase our taxonomic understanding of such important genera such as *Anthophora*, *Eucera*, *Hoplitis*, *Anthocopa*, *Megachile* and *Chalicodoma*.
4) To establish a group of researchers with a shared interest in the pollination biology of the Mediterranean Basin, under the auspices of MEDECOS and to promote and maintain multinational, perennial projects; to establish links with biologists working along similar lines in other regions with a Mediterranean-type climate and biota.

We regard such projects as vital to the future of the Mediterranean biota. Much of the native flora is under threat from overgrazing, tourist development and increasingly intensive agriculture. If we are to formulate viable conservation policies and management plans, then we must understand the dynamics of pollination at the community level. This must include investment in improving our knowledge of bee systematics. This is vital not only for theoretical considerations of pollinator-plant relationships, but has potential economic value. We must regard our wild bee faunas as reserves of potential pollinators which may be suitable for management for crop pollination (O'Toole, 1993), especially in areas such as Spain and Israel, where the cultivation of exotic fruits and vegetables is increasing.

Acknowledgements

We thank Prof. Avi Shmida of the Hebrew University, Jerusalem and Prof. R. Menzel of Free University of Berlin for permission to quote their unpublished work on the colour reflectance of flowers. We thank Dr. G.C. McGavin for his critical reading of this paper in manuscript.

References

Baker, H.G. & Hurd, P.D. 1968. Intraflora ecology. Ann. Rev. Ent. 13: 385–414.

Bawa, S.K. 1983. Patterns of flowering in tropical plants. In: Jones, C.E. & Little, R.J. (eds.), A Handbook of Experimental Pollination Ecology. Van Nostrand, New York, pp. 394–410.

Cohen, D. & Shmida, A. 1993. The evolution of flower display and reward. Evol. Biol. 68: 81–120.

Corbet, S.A., Williams, I.H. & Osbourne, J.L. 1991. Bees and the pollination of crops and wild flowers: changes in the European Community. A review commissioned by Scientific and Technical Options Assessment, European Parliament.

Dafni, A. 1983. *Orchis caspia* (Orchidaceae) – nectarless species deceiving the pollinators of nectariferous species from other families. J. Ecol. 71: 467–474.

Dafni, A. 1984. Mimicry and deception in pollination. Ann. Rev. Ecol. Syst. 15: 259–273.

Dafni, A. 1986. The Mediterranean maquis as a pollination environment. 5th OPTIMA meeting. Istanbul.

Dafni, A. 1987. Pollination in *Orchis* and related genera: evolution from reward to deception. In: Arditti J. (ed.),. Orchid Biology: Reviews and Perspectives, IV. Comstock Publ. Associates, Cornell University Press, Ithaca and London, pp. 79–104.

Dafni, A. 1990. Advertisement, flower longevity, reward and nectar protection in Labiatae. Act. Hort. 288: 340–346.

Dafni, A. 1991. The reaction of *Amphicoma* spp. beetles to red flower models and its implications in pollination. Plant Reproductive Ecology: Progress and Perspectives. Symposium, Uppsala, Sept. 1991 (Abstract).

Dafni, A. & Bernhardt, P. 1990. Pollination of terrestrial orchids of southern Australia and the Mediterranean region. Evol. Biol. 24: 193–252.

Dafni, A. & Werker, E. 1982. Pollination ecology of *Sternbergia clusiana* (Ker–Gawler) Sprng. (Amaryllidaceae). New Phytol. 91: 571–577.

Dafni, A., Bernhardt, P., Shmida, A., Ivri, Y., Greenbaum, S., O'Toole & Losito, L., 1990. Red bowl-shaped flowers: convergence for beetle pollination in the Mediterranean region. Isr. J. Bot. 89: 31–42.

Dukas R. & Shmida, A. 1990. Progressive reduction in the mean body sizes of solitary bees active during the flowering season and its correlation with the sizes of the bee flowers of the mint family (Lamiaceae). Isr. J. Bot. 133–141.

Faegri, K. & Van der Pijl, L. 1979. Principles of Pollination Ecology. 3rd. Edition. Pergamon Press, Oxford.

Frankie, G.W., Haber, W.W., Opler, P.A. & Bawa, K.S. 1983. Characteristics and organization of the large bee pollination system in the Costa Rican dry forest. In: Jones, C.E. & Little, R.J. (eds.) Handbook of Experimental Pollination Biology. Scientific and Academic Editions, Van Nostrand, New York, pp. 441–448.

Grebbenikov, V.S. 1972. [On the question of preserving bumblebees, important pollinators.] In: Okhrana prirody i ratsional'noe ispol'zovanie prirodynkh resursov tsentral'nochernozemnoi polosy. Ed. Skuf'in, K.V. Voronezh, USSR; Voronezhskii Ordena Lenina Gosudarstvennyi Universitet Imeni Leninskogo Komsomola. [In Russian, seen in abstract.], pp. 78–82.

Herrera, C.M. 1982. Seasonal variation in the quality of fruits and diffuse coevolution between plants and avian dispersers. Ecology 63: 773–775.

Herrera, C.M. 1987. Components of pollinator 'quality': comparative analysis of a diverse insect assemblage. Oikos 50: 79–90.

Herrera, J. 1985. Nectar secretion patterns in southern Spanish Mediterranean shrublands. Isr. J. Bot. 34: 47–58.

Herrera, J. 1986. Flowering and fruiting phenology in the coastal shrublands of Doñana, South Spain. Vegetatio 68: 91–98.

Herrera, J. 1987. Flower and fruit biology in southern Spanish Mediterranean shrublands. Ann. Mo. Bot. Gdn. 74: 69–78.

Herrera, J. 1991. The reproductive biology of a riparian Mediterranean shrub, *Nerium oleander* L. (Apocynaceae). Bot. J. Linn. Soc. 106: 147–172.

Kullenberg, B. & Bergström, G. 1976. Hymenoptera Aculeata males are pollinators of *Ophrys* orchids. Zool. Scr. 5: 13–23.

Kummerov, J. 1983. Comparative phenology of Mediterranean-type plant communities. In: Kruger, F.J., Mitchell, D.T. & Jarvis, J.U.M. (eds.). Mediterranean-type Ecosystems: The Role of Nutrients. No. 43, Springer, Berlin, pp. 300–317.

LaBerge, W.E. 1991. The Bees of Mexico. Melissa 4: 1–2.

Menzel, R. & Shmida, A. 1993. The ecology of flower colour and the natural colour vision system of insect pollinators. Biol. Rev. 68: 81–120.

Michener, C.D., 1979. Biogeography of the bees. Ann. Mo. Bot. Gdn. 66: 277–347.

O'Toole, C. 1991. Wild bees, systematics and the pollination market in Israel. Antenna 15: 66–72.

O'Toole, C. 1993. Diversity of native bees and agroecosystems. In: LaSalle, J. & Gauld, I. (eds.). Hymenoptera and Biodiversity. Symposium of the Third Quadrennial Congress of the International Society of Hymenopterists.

Commonwealth Agricultural Bureau International, London, pp. 69–106.
O'Toole, C. & Raw, A. 1991. Bees of the World. Blandford Press, London.
Paulus, H.H. & Gack, C. 1990. Pollinators as pre-pollinating isolation factors: evolution and speciation in *Ophrys* (Orchidaceae). Isr. J. Bot. 39: 43–80.
Petanidou, T. & Vokou, D. 1990. Pollination and pollen energetics in Mediterranean ecosystems. Am. J. Bot. 77: 986–992.
Pignatti, S. 1978. Evolutionary trends in Mediterranean flora and vegetation. Vegetatio 37: 175–185.
Quezel, P. 1985. Definition of the Mediterranean region and the origin of the flora. In: Gomez-Campo, C. (ed.), Plant Conservation In The Mediterranean Area. Dr W. Junk Publishers, Dordrecht, pp. 9–43.
Roubik, D.W. 1988. An overview of Africanized honeybee populations: reproduction, diet and competition. In: Needham, M., Delfinado-Baker, R., Page, R. & Bowman, C. (eds.). Proceedings of the International Conference on Africanized Honeybees and Bee Mites. E. Horwood, Ltd., Chichester, England, pp. 45–54.
Selten, R. & Shmida, A. 1991. Pollinator foraging and flower competition in a game equilibrium model. In: Selten, R. (ed.). Game and Equilibrium Models Theory, Vol. 1. Springer, Berlin, pp. 195–256.
Shmida, A. 1981. Mediterranean vegetation of Israel and California, similarities and differences. Isr. J. Bot. 30: 105–123.
Shmida, A. & Dafni, A. 1989. Blooming strategies, flower size and advertising in the 'lily-group' geophytes in Israel. Herbertia 45: 111–123.
Stebbins, G.L. 1970. Adaptive radiation of reproductive characteristics in angiosperms. I. Pollination mechanisms. Ann. Rev. Ecol. Syst. 1: 307–326.
Vogel, S. & Westerkamp, C. 1991. Pollination: an integrating factor of biocenoses. In: Seitz, A. & Loeschcke, V. (eds.). Species Conservation: A population-biological approach. Birkhäuser Verlag, Basel, pp. 159–170.
Weevers, T. 1952. Flower colours and their frequency. Acta Bot. Neerl. 1: 81–92.
Williams, P. 1989. Bumblebees and their decline in Britain. Central Association of Beekeepers. June 1989: 1–15.

CHAPTER 13

Red flowers and butterfly pollination in the fynbos of South Africa

S.D. JOHNSON and W.J. BOND
Department of Botany, University of Cape Town, Rondebosch 7700, South Africa

Key words: convergent evolution, fynbos, pollination, psychophily, sunbirds, nectar, *Meneris tulbaghia*

Abstract. A guild of fynbos species with red flowers is pollinated exclusively by the butterfly *Meneris tulbaghia*. Such dependence on a single species of pollinator is rarely found in plants. Species pollinated by *M. tulbaghia* share several convergent characteristics including large, red flowers with straight, narrow nectar tubes and a flowering period in late summer. The butterfly appears to be attracted primarily to the red colour of the flowers as shown by experiments in which butterflies chose red coloured model flowers over other colours. The straight, narrow nectar tubes of flowers in the guild discourage visits by sunbirds which otherwise visit red flowers. Nectar properties of plants in the guild vary considerably, but most species have nectar of between 15 and 25% sugar concentration. Similar low sugar concentrations were found in a sample of bird-pollinated fynbos species, many of which are regularly 'robbed' of nectar by the butterfly. Perception of red colour by *M. tulbaghia* may have evolved because it facilitated exploitation of the copious amounts of dilute nectar available in most red bird-pollinated flowers. Plants pollinated by other insects may have shifted to *M. tulbaghia* because of the efficiency of a pollinator which visits flowers of only one colour. Since about 15 of the most spectacular fynbos species depend solely on *M. tulbaghia* for pollination, the butterfly can be considered a 'keystone' species with a high priority for conservation.

Introduction

The association between red flowers and pollination by nectarivorous birds is well known (Grant, 1966; Raven, 1973). Insect-pollinated plants on the other hand rarely have red flowers. For example, the near absence of bird pollination in Europe is generally considered the reason for the paucity of red flowers in the European flora. Perception of red flowers, however, is not entirely restricted to birds. In Israel a small guild of red-flowered plants is pollinated by *Amphicoma* beetles (Dafni *et al.*, 1990). Some flowers in tropical America with contrasting red and yellow patterns (e.g. *Lantana camara*, *Caesalpinia pulcherrima*) are visited by butterflies (Cruden & Hermann-Parker, 1979; Boyden, 1980). Butterflies may have the widest visual spectrum of all animals, with perception of both ultraviolet and red light in some species (Bernard, 1979; Eguchi *et al.*, 1982).

In South Africa, a number of red-flowered species are pollinated by the butterfly *Meneris* (*Aeropetes*) *tulbaghia* (hereafter simply termed *Meneris* as the genus is monotypic). It is outstanding among the Satyridae for its large size (wingspan = 80 mm), swift flight and flower-visiting habits. *Meneris* is found throughout the moister mountain ranges of Southern Africa and favours rocky outcrops and gorges (Johnson & Bond, 1992). It is fairly common in the fynbos, a region relatively depauperate in butterflies (Cottrell, 1985).

The attraction of the butterfly to red flowers was noted by early lepidopterists (Trimen, 1887) and botanists (Marloth, 1896, 1915). Marloth (1896) showed that the spectacular red flowers of the orchid *Disa uniflora* are pollinated by *Meneris*, which carries the pollinia on its legs. The butterfly is also well known to mountaineers for its habit of swooping down on red hats and socks! Despite popular interest in this unusual butterfly (Johnson, 1992b), no proper study of its pollination relationships had been undertaken. This study was based on the following questions:

1. Which species are pollinated by *Meneris*? Although Marloth had recorded visits to many

M. Arianoutsou and R.H. Groves, Plant-Animal Interactions in Mediterranean-Type Ecosystems, 137–148, 1994.

species with red flowers, many other inaccessible species had not been examined in the field.
2. Do these species pollinated by *Meneris* share convergent characters, e.g. colour, morphology, flowering time and nectar properties?
3. Can these convergent characters (if any) be explained by the peculiarities in the foraging behaviour, phenology and morphology of the butterfly?

Methods

Field observations

Field work took place at several sites in the western, southern and eastern Cape Province during 1990–1992. Plants known, or suspected, to be visited by *M. tulbaghia* were observed in the field. Careful attention was paid to whether the butterfly contacted the anthers and stigma of flowers and whether pollen was present on the body of the insect.

Colour analysis

The spectral reflectances of flowers of each species were measured using an ACS 550 nm spectrophotometer. This method avoids the problem of subjectively assessing colour. Only the attractive parts of the flower were measured. The reflectance curve from 400–700 nm gives an objective representation of colour in the wavelengths visible to most butterflies. In addition the spectra were analyzed using segment classification (Endler, 1990). This involves dividing the spectrum into four equal segments, finding the differences between the areas under the curve at alternating segments and plotting the two 'differences' on a two- dimensional colour space. This method allows comparison between a number of reflectance curves at once and compensates for differences in brightness. It approximates signal processing in animal colour vision which is thought to be based on opponency between signals from photoreceptors with different spectral sensitivities (Endler, 1990). Some butterflies are thought to have up to four visual pigments (Bernard, 1979), as opposed to humans and bees which have only three. Spectral reflectances from a sample of bird pollinated plants were also analyzed.

The colour preference of *Meneris* was examined experimentally by recording the relative proportions of visits to various coloured paper discs (model flowers). These were displayed randomly either in arrays in the field (1990) or in outdoor flight cages containing captive butterflies (1991). A reward of 20% honey solution was offered in glass funnels mounted in the centre of the paper discs in 1991. We also recorded the proportion of visits to red paper discs of different sizes displayed in the field during 1990.

Nectar analysis

Nectar volumes of flowers in the field were measured using microcapillaries. A Bellingham and Stanley 0–50% refractometer was used to determine nectar sugar concentration. Nectar was extracted from flowers as close to 1200 h as possible in order to standardise results and because butterfly activity is high at this time. Nectar was also spotted onto Whatman #1 filter paper and kept refrigerated until later analysis of sugar constituents using high pressure liquid chromatography (HPLC), could be performed.

Results

Field observations

Through field observations and examining butterflies for pollen it became clear that fynbos species with red flowers which have nectar tubes which are narrow, straight and vertical were invariably pollinated by *Meneris*. These species all flower between December and April, corresponding to the flight period of the butterfly (Johnson, 1992a). The narrow nectar tubes of these species appear to prevent sunbirds from gaining access to the nectar. Sunbirds can obtain nectar from *Crassula coccinea* only by piercing the narrow nectar tube. I observed a sunbird which apparently managed to extract nectar from *Nerine sarniensis*, but the bird did not contact the anthers due to the brush-type morphology of this flower which is adapted to *Meneris* (see discussion). By contrast, the narrow proboscis of *Meneris* allows it to frequently

Table 13.1. Species conforming to the syndrome of pollination by *Meneris tulbaghia*

Family/species	Study site	Flower type	Flowering time
Crassulaceae			
Crassula coccinea	Table Mountain	Classic	Dec-Feb
Amaryllidaceae			
Brunsvigia marginata	Du Toits Kloof, Bains Kloof	Brush	Feb-Mar
Cyrtanthus elatus	Robinson Pass	Brush	Dec-Feb
C. guthrieae	Bredasdorp reserve	Brush	March
*C. montanus**		Brush	Jan-Feb
Nerine sarniensis	Bettys Bay	Brush	Mar-Apr
Iridaceae			
Gladiolus cardinalis	Du Toits Kloof, Bains Kloof	Flag	Dec-Jan
G. stefaniae	Montagu	Flag	Mar-Apr
G. nerinoides!	Jonkershoek	Classic	Jan-Feb
G. sempervirens	George Peak	Flag	Mar
G. cruentus #$		Flag	Jan
G. saundersoniae #$		Flag	Feb-Mar
*G. stokei**		Flag/classic	Mar
*Tritoniopsis leslei**		Classic	Feb-Mar
*T. longituba**		Horizontal	Dec-Mar
Schizostylis coccinea $	Maclear district	Brush	Jan-Mar
Orchidaceae			
Disa uniflora	Table Mountain	Flag	Dec-Feb
Disa ferruginea	Table Mountain	Horizontal	Feb-Mar
*D. porrecta** $		Horizontal	Feb

* Species which conform to the *Meneris* syndrome but which have not been observed in the field.
! Visitation observed by C.L.Wicht (letter housed in Bolus Herbarium).
Visitation by *Meneris* observed by J. Vlok (pers. comm.).
$ Species occurring outside the fynbos region.

'rob' nectar from bird-pollinated flowers. Pollination by *Meneris* occurs in at least four families and eight genera (Table 13.1). Nineteen South African species, of which fifteen are endemic to the fynbos, appear to rely solely on *Meneris* for pollination. All these species, with the exception of *Crassula coccinea* (Crassulaceae), are petaloid monocotyledons.

The floral morphology varies greatly within the guild. We identify four floral types: *Classic-type* flowers (see Faegri & Van der Pijl, 1979) have a narrow upright tube with inserted stamens and a flat rim (Fig 13.1 b-d). *Brush-type* flowers have tall extended stamens projecting from a funnel-shaped flower (Fig 13.2 a-d). *Flag-type* flowers are large and showy with a broad landing surface (Fig 13.3 a-d). *Horizontal-type* flowers resemble bird-pollinated flowers closely, except that in some cases the tube is too narrow to allow access of a bird's bill.

Colour analysis

Without exception, species pollinated by *Meneris* have spectral reflectances with a maximum positive slope between 570 and 650 nm, a curve shape perceived as red by humans (Fig 13.4). If these curves, as well as curves for bird-adapted flowers, are analyzed by segment classification, it can be seen that no substantial difference exists in the colours of *Meneris* and many bird-adapted flowers (Fig 13.5). The strong preference shown for red-coloured model flowers (Fig 13.6) is a striking feature of the behaviour of *Meneris*. Large model flowers were also strongly preferred over smaller models (Fig 13.6).

Nectar analysis

The nectar properties of plants adapted to *Meneris* showed great variation in volume, concentra-

Fig. 13.1. (a) *Chasmanthe floribunda*, a plant adapted for pollination by sunbirds. Note the curved, horizontal perianth tube and the overarching anthers positioned to place pollen on the head of the sunbird; (b) *Gladiolus nerinoides*, an example of a classical butterfly-pollinated flower with a straight, upright perianth tube and inserted anthers; (c) *Meneris tulbaghia* feeding on *Crassula coccinea*; (d) Cross-section of a flower of *C. coccinea* showing the straight perianth tube and inserted anthers which place pollen on the proboscis of *Meneris*.

Fig. 13.2. (a) *Cyrtanthus guthrieae*, a brush-type flower with highly exerted stamens; (b) *Meneris tulbaghia* on *Cyrtanthus elatus*; (c) *Schizostylis coccinea*, a brush-type flower; (d) *Brunsvigia marginata*, a brush-type flower.

Fig. 13.3. (a) The flag-type flowers of *Disa uniflora* are among the largest known in orchids; (b) When *Meneris* settles on the flower of *D. uniflora* pollinia are 'glued' to its legs and thorax; (c) *Gladiolus cardinalis*, a flag type flower; (d) *Meneris* feeding on *G. cardinalis*.

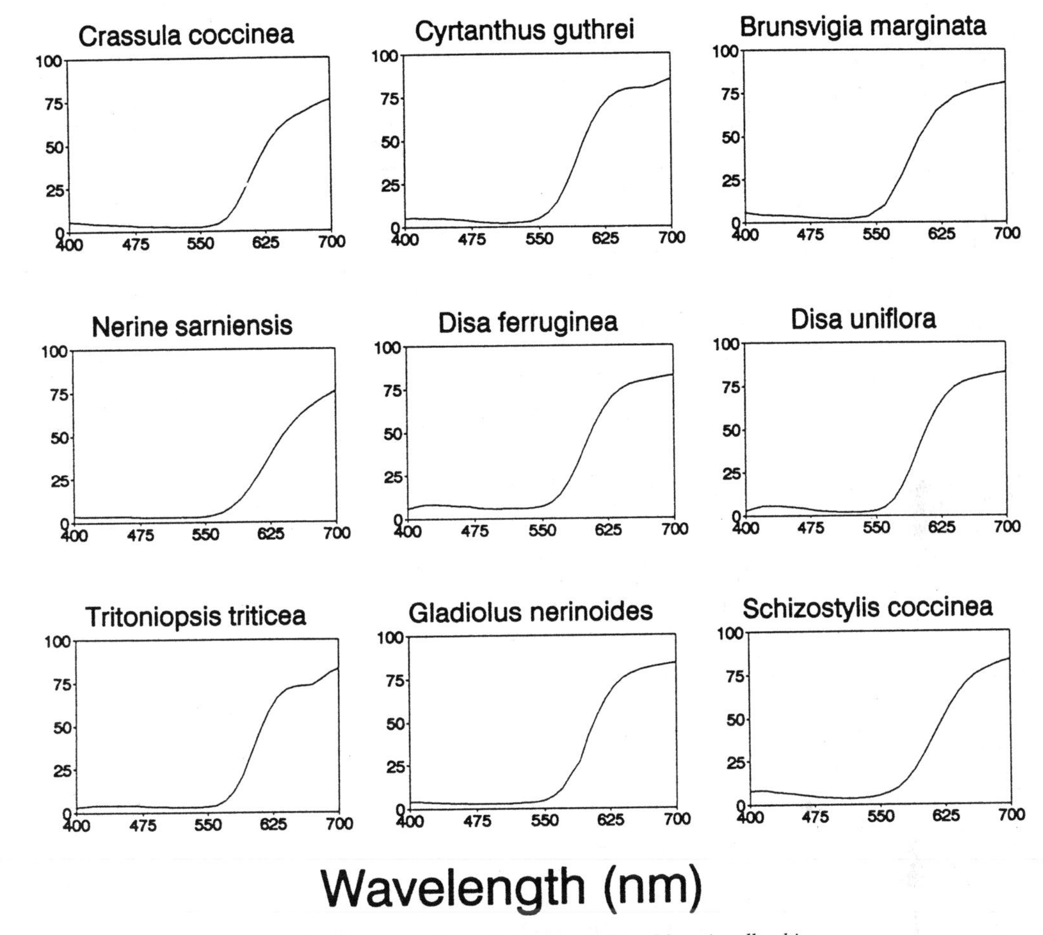

Fig. 13.4. Spectral reflectances of species pollinated by *Meneris tulbaghia*.

tion and sugar proportions (Table 13.2). Comparison with a sample of bird-adapted nectars shows that the average nectar volumes and concentrations are almost identical (Tables 13.2, 13.3). The orchid *Disa ferruginea* does not produce nectar and appears to mimic a sympatric nectar-producing species *Tritoniopsis triticea* (unpublished data).

Analysis of the constituent nectar sugars did not reveal the trend towards sucrose dominated nectar in butterfly pollinated plants that has been reported elsewhere (Cruden & Hermann-Parker, 1979; Baker & Baker, 1983). Four species in the sample of *Meneris* pollinated plants have sucrose dominated nectar and four species have nectar dominated by hexose sugars (glucose and fructose). Among the sample of species adapted for bird pollination, seven have sucrose dominated nectar and four have hexose dominated nectar.

Discussion

A unique pollination syndrome?

A pollination syndrome is a set of convergent characteristics that appears in unrelated plant groups and which is associated with the peculiar selective pressures imposed by shared pollinators. We recognise the following convergent characteristics in *Meneris* pollinated plants: red flowers, large advertising area, dilute nectar, narrow, straight perianth tubes often with an upright orientation, and a flowering season from December to April. This '*Meneris* syndrome' combines elements of both the classical bird and butterfly syndromes. The red flowers are typical of bird-pollinated flowers, but birds are excluded by the floral morphology. The floral morphology in some species is typical of other butterfly-polli-

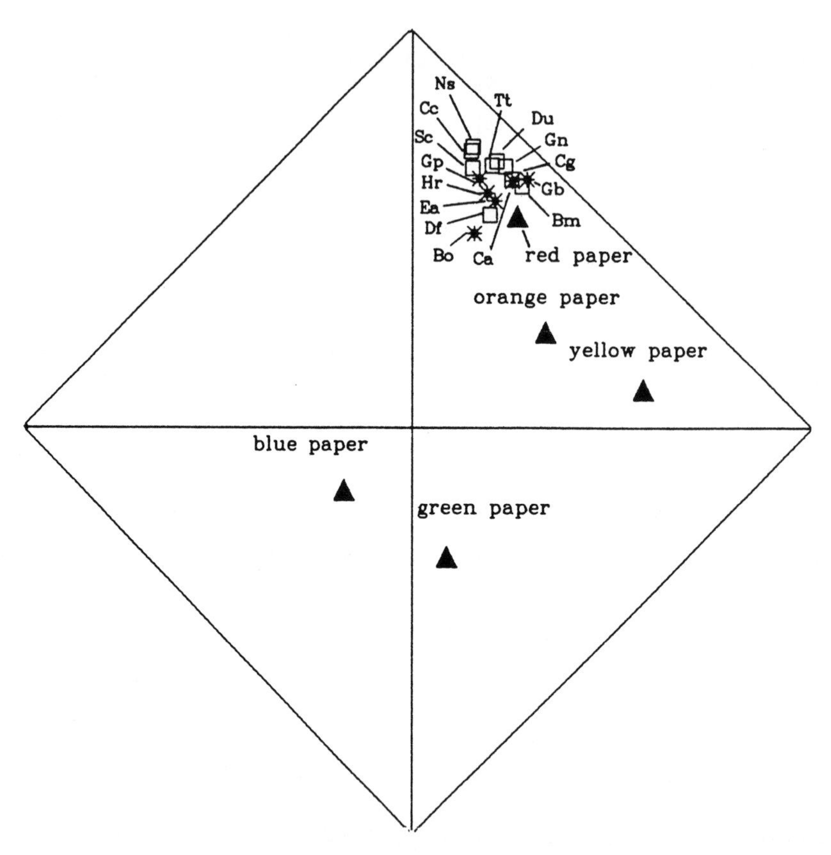

Fig. 13.5. Segment classification (Endler, 1990) of reflectance spectra of *Meneris*-pollinated plants (□), bird pollinated plants (*) and papers used in the discrimination experiments in 1991 (▲). The position of colours in the colour space is independent of human perception. Spectra closer to the origin are less saturated (paler, less chroma) and the angle with respect to the origin is a measure of the shape of the spectrum or hue. Species abbreviations are as follows. Ns – *Nerine sarniensis*, Tt – *Tritoniopsis triticea*, Du – *Disa uniflora*, Gn – *Gladiolus nerinoides*, Cg – *Cyrtanthus guthrieae*, Gb – *Gladiolus bonaespei*, Bm – *Brunsvigia marginata*, Ca – *Chasmanthe aethiopica*, B0 – *Brunsvigia orientalis*, Df – *Disa ferruginea*, Ea – *Erica abietina*, Gp – *Gladiolus priorii*, Hr – *Haemanthus rotundifolius*, Sc – *Schizostylis coccinea*, Cc – *Crassula coccinea*.

nated plants, but the coloration appears to exclude other butterflies. For example, *Princeps demodocus* (Papilionidae) is a large butterfly which, while often sympatric with *Meneris*, appears to ignore the flowers chosen by *Meneris* and instead tends to visit blue flowers. Thus by virtue of excluding birds (through morphology) and other insects (through colour), *Meneris*-pollinated plants rely almost exclusively on this single insect for pollination.

The various floral morphologies found in *Meneris*-pollinated plants represent distinct strategies for placing pollen on the butterfly. Classic-type flowers, with a narrow, upright nectar tube and flat rim, have inserted or partially inserted anthers which place pollen on the head and proboscis of the butterfly, e.g. *Crassula coccinea* (Fig 13.1 c,d). Brush-type flowers (Fig 13.2 a-d) are most interesting as they represent the biggest deviation from the general butterfly-pollination syndrome. These flowers have straight, narrow, upright nectar tubes, but the anthers and stigma are highly exserted and function to brush pollen onto the wings and bodies of flying butterflies. Brush-type flowers are effective as they exploit a habit of *Meneris* of swooping down on flowers without landing. During these '*inspection*' visits, which usually outnumber feeding visits, pollen is brushed onto the butterfly. In *Brunsvigia marginata* and *Nerine sarniensis* the butterfly does not even contact the anthers during a feeding visit and pollination takes place almost exclusively during

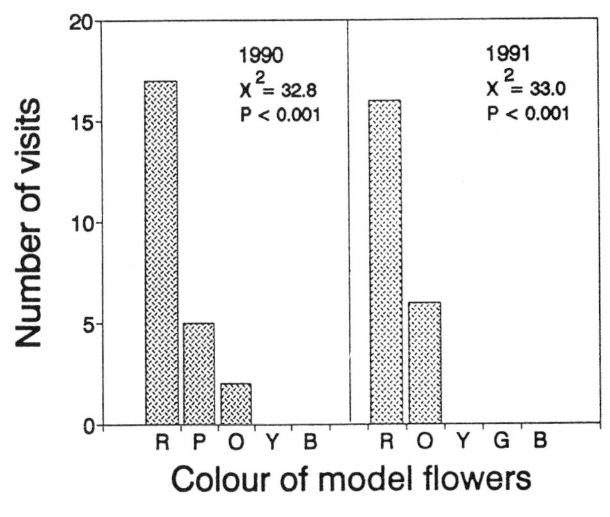

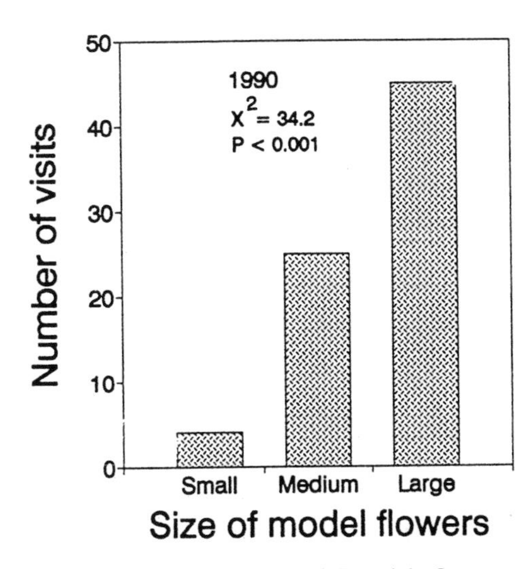

Fig. 13.6. Discrimination by *Meneris tulbaghia* between flower models differing in colour and size. R = red, P = pink, O = orange, Y = yellow, G = green, B = blue. The diameters of the paper discs were as follows: small = 5 cm, medium = 10 cm, large = 15 cm.

inspection visits. Inspection visits appear to form a higher proportion of the butterfly's daily activity later in the season and may be associated with territoriality and mate location. The wings and body are used for pollen placement as opposed to the proboscis which is considered a poor pollen-transporting surface (Wiklund *et al.*, 1979).

Flag-type flowers, e.g. *Disa uniflora*, have a large advertising area and a broad landing surface (Fig 13.3). The site of pollen placement varies from the butterfly's legs, as in *Disa uniflora*, to its wings and body, as in *Gladiolus cardinalis*. Finally, horizontal-type flowers are closely allied to bird pollination and in species such as *Tritoniopsis triticea*, pollination by *Meneris* appears to be facultative to bird pollination. Examination of the stigmas of *Tritoniopsis triticea* showed that the majority were plastered with wing scales of *Meneris*, suggesting that pollen transfer may take place when the wings of a feeding butterfly contact the protruding anthers and stigmas of surrounding flowers on the inflorescence. In *Disa*

Table 13.2. Nectar properties of species adapted to pollination by *Meneris tulbaghia*

Family/species	Volume		Concentration		Sugars (%)		
	μl	n	g/100 g	n	Fru	Glu	Suc
Crassulaceae							
Crassula coccinea	2.4	(8)	16.4	(8)	54	39	7
Amaryllidaceae							
Brunsvigia marginata	3.1	(26)	26.5	(15)	34	44	22
Cyrtanthus elatus	–		–		20	26	54
C. guthrei	9.8	(11)	17.7	(9)	14	19	67
Nerine sarniensis	4.3	(10)	36.0	(8)	50	42	8
Iridaceae							
Gladiolus cardinalis	9.4	(12)	24.8	(12)	9	16	75
G. stefaniae	–		–		10	12	78
Schizostylous coccinea	1.9	(19)	17.4	(11)	42	43	15
Orchidaceae							
Disa ferruginea	No nectar produced						
D. uniflora	40.9	(5)	8.0	(5)	0	0	100
Average	10.2		21.0		29	30	41

Table 13.3. Nectar properties of species adapted to pollination by birds

Family/species	Volume µl	n	Concentration g/100 g	n	Sugars (%) Fru	Glu	Suc
Amaryllidaceae							
Brunsvigia orientalis	13.2	(3)	34.8	(3)	18	25	58
*Cyrtanthus ventricosus**	1.1	(10)	15.2	(9)	–	–	–
*Haemanthus coccinea**	2.0	(23)	33.0	(7)	–	–	–
H. rotundifolius	5.5	(16)	23.8	(14)	43	54	3
Iridaceae							
Chasmanthe aethiopicum	10.1	(6)	13.9	(6)	46	51	3
C. floribunda	15.7	(8)	16.8	(8)	45	46	9
Gladiolus priorii	13.4	(4)	26.3	(3)	6	13	81
G. watsonius	15.6	(12)	27.9	(12)	1	2	97
*Tritoniopsis burchelli**	6.2	(9)	20.8	(8)	16	16	69
T. nervosa	17.1	(7)	33.7	(10)	6	8	86
*T. triticea**	2.3	(7)	24.2	(6)	9	10	81
*Watsonia tabularis**	25.6	(5)	14.5	(4)	17	21	62
Witsenia maura	27.8	(10)	13.5	(10	45	55	0
Average	12.0		23.0		23	27	50

* Species which are important sources of nectar to *Meneris tulbaghia* and which may be partially pollinated by the butterfly.

ferruginea which has narrow horizontal spurs, pollination is exclusively by *Meneris* which carries pollinia attached to the proboscis. *Disa ferruginea* has often been considered, on the basis of morphological evidence only, to be bird-pollinated (e.g. Vogel, 1954).

In terms of nectar properties, there is no substantial difference between species adapted to *Meneris* and species adapted to birds (Tables 13.2, 13.3). Much of the nectar requirements of *Meneris* are met by bird-pollinated plants. In the summer rainfall area of South Africa, *Kniphofia* (Asphodelaceae), an ornithophilous genus, appears to be the primary nectar source for *Meneris*. In the fynbos, bird-pollinated members of *Tritoniopsis*, *Watsonia* (both Iridaceae), *Kniphofia* (Asphodelaceae) and *Erica* (Ericaceae) are all important nectar sources for the butterfly. The dilute nectar of bird-adapted flowers is ideal for butterflies which have to suck nectar through a narrow proboscis. The widely differing proportions of sugars making up the nectar of both *Meneris* and sunbird pollinated plants (Tables 13.2, 13.3) suggests that sugar ratios play little role in nectar selection by these pollinators. Sugar ratios in the nectar appear to reflect the phylogenetic affinities of taxa (cf. Percival, 1961), rather than adaptation to *Meneris* or sunbirds.

Evolution of the guild

Although red coloration is known in butterfly-pollinated flowers elsewhere (Cruden & Herman-Parker, 1979; Boyden, 1980), the existence of a guild with entirely red flowers dependent on a single butterfly for pollination must be unique. Two questions are immediately raised. Firstly, what advantage could *Meneris* have in possessing such a distinct preference for red? Secondly, why have unrelated plant species converged on this butterfly for pollination?

A spontaneous colour preference has been reported for many butterflies. Swihart & Swihart (1970), for instance, found that *Heliconius charitonius* had a spontaneous preference for blue-green and orange-red model flowers, but that they could be trained to feed on other colours. Preference by butterflies for blue and yellow flowers was noted by Ilse (1928, in Swihart & Swihart, 1970). In *Pieris brassicae* (Pieridae) uncoiling of the proboscis in preparation for feeding occurs in response to blue and orange-red light (Scherer & Kolb, 1987). Distinct colour preferences of various butterfly species feeding on the pink and orange flowers of *Lantana camara* was reported by Dronamraju (1960).

What advantage is red perception to a butterfly? In some butterflies with red wing patterns, discrimination of red assists in mating behaviour. However, there is no red in the wing patterns of *Meneris* and preference for the colour seems to be associated purely with feeding. In most plant communities red flowers signal the presence of the copious nectar found in bird-adapted flowers. It has been suggested that this nectar resource is disguised from most insects by virtue of their poor perception of red (at least in some bees) (Raven, 1973). However, *Meneris* obtains a large percentage of its nectar requirements from bird-adapted flowers, most of which have red coloured flowers. The evolution or inheritance of red perception in *Meneris* must have therefore facilitated exploitation of this abundant nectar resource. Butterflies are particularly capable of 'robbing' nectar from flowers adapted to other pollinators and it has been suggested that butterflies evolved long proboscides to enable them to exploit a wide range of flower morphologies (Wiklund *et al.*, 1979). The dilute nectar found in bird-pollinated plants is especially suitable for butterflies – nectar which is too viscous cannot be sucked by butterflies (Kingsolver & Daniel, 1979). We suggest, therefore, that through its ability to perceive red colours, *Meneris* evolved a parasitic lifestyle of robbing bird-pollinated flowers.

The effectiveness of *Meneris* as a pollinator is probably directly related to its narrow colour preference. A specialised pollinator holds many advantages. Since very few flowers in any community have red flowers, plants adapted to *Meneris* are assured that a minimum of pollen is wasted on foreign stigmas and conversely that a minimum of foreign pollen clogs its own stigmas. Perhaps more important, however, is the increased probability of visitation because of the 'magnetic' attraction of the butterfly to red flowers. This may be of particular benefit to the rare plants in the guild which are quickly 'noticed' by the butterfly, despite being a small minority of the nectar-producing plants in a community. A feature of the guild is the large number of rare species. Four species (*Gladiolus nerinoides*, *G. sempervirens*, *G. stefaniae* and *Cyrtanthus guthrei*) are in the Red Data Book for fynbos (Hall & Veldhuis, 1985) and many others have very restricted distributions.

A wide range of unrelated plants have converged to the syndrome of *Meneris* pollination. Many species pollinated by other butterflies and long proboscid flies are morphologically pre-adapted to *Meneris*; all that is required for a shift to *Meneris* pollination is a mutation for red coloration. Dronamraju (1960) has already pointed out that the strict colour preferences of some butterflies might allow a sympatric speciation event through a single colour mutation.

Conclusion

Perhaps the most striking feature of this pollination guild is the near total dependence of the species on a single insect for pollination by virtue of both colour and morphological 'filters' that exclude other insects and birds respectively. The survival of these plants is thus directly linked to conservation of the butterfly. Declining populations of Satyrid butterflies on the Cape Peninsula have been attributed to overly frequent wildfires (Claassens & Dickson, 1980). The conservation of these spectacular red-flowered species probably rests on the correct fire-management of fynbos.

Acknowledgements

SDJ is grateful to Kathy Johnson for help in the field. Ben-Erik van Wyk kindly analyzed the nectar sugar proportions. Plascon Paint company kindly made their spectrophotometer available for spectral analyses of flowers. This work was financially supported by the FRD Special Programme for Evolutionary Biology and the Charles Dixon Award.

References

Bernard, G.D. 1979. Red-absorbing visual pigment of butterflies. Science 203:202–204.

Baker, H.G. & Baker, I. 1983. A brief historical review of the chemistry of floral nectar. In: Bentley, B. & Elias, T. (eds). The Biology of Nectaries. Columbia University Press, New York, pp. 126–152.

Boyden, T.C. 1980. Floral mimicry by *Epidendrum ibaguense* (Orchidaceae) in Panama. Evolution 34:135–136.

Claassens, A.J.M. & Dickson, C.G.C. 1980. The butterflies

of the Table Mountain range. C. Struik Publishers, Cape Town.

Cottrell, C.B. 1985. The absence of coevolutionary associations with Capensis floral elements in the larval/plant relationships of southwestern Cape butterflies. In: Vrba, E.S. (ed). Species and speciation. Transvaal Museum Monograph No. 4, pp. 155–124.

Cruden, R. W. & Hermann-Parker, S. M. 1979. Butterfly pollination of *Caesalpinia pulcherrima*, with observations on a psychophilous syndrome. J. Ecol. 67:155–168.

Dafni, A., Bernhardt, P., Shmida, A., Ivri, Y., Greenbaum, S., O'Toole, C. & Losito, L. 1990. Red bowl-shaped flowers: convergence for beetle pollination in the Mediterranean region. Israel. J. Bot. 39:81–92.

Dronamraju, K.R. 1960. Selective visits of butterflies to flowers: a possible factor in sympatric speciation. Nature 186:178.

Eguchi, E., Watanabe, K., Hariyama, T. & Yamahoto, K. 1982. A comparison of electrophysiologically determined spectral responses in 35 species of Lepidoptera. J. Insect Physiol. 28:675–682.

Endler, J.A. 1990. On the measurement and classification of colour in studies of animal colour patterns. Biol. J. Linn. Soc. 41:315–354.

Faegri, K. & van der Pijl, L. 1979. The principles of pollination ecology. 3rd edition. Pergamon, Oxford.

Grant, K.A. 1966. A hypothesis concerning the prevalence of red coloration in California hummingbird flowers. Amer. Nat. 100:85–97.

Hall, A.V. & Veldhuis, H.A. 1985. South African red data book: plants – Fynbos and Karoo biomes. South African National Scientific Programmes Report no 117, CSIR, Pretoria.

Johnson, S.D. 1992a. Plant animal relationships. In: Cowling, R.M. (ed). The ecology of fynbos: fire, nutrients and diversity. Oxford University Press, Cape Town. pp. 175–205.

Johnson, S.D. 1992b. A butterfly with a passion for red. Afr. Wildlife 46:176–179.

Johnson, S.D. & Bond, W.J. 1992. Habitat dependent pollination success in a Cape orchid. Oecologia 91:455–456.

Kingsolver, J.G. & Daniel, T.L. 1979. On the mechanics and energetics of nectar feeding in butterflies. J. Theor. Biol. 76:167–179.

Marloth, R. 1896. The fertilization of '*Disa uniflora*' Berg, by insects. Trans. S. Afr. Phil. Soc. 8:93–95.

Marloth, R. 1915. The Flora of South Africa. Vol 4. Monocotyledones. Darter Bros & Sons, London.

Percival, M.S 1961. Types of nectar in angiosperms. New Phytol. 60:235–281.

Raven, P.H. 1973. Why are bird-visited flowers predominantly red? Evolution 26:674.

Scherer, C. & Kolb, G. 1987. Behavioral experiments on the visual processing of color stimuli in *Pieris brassicae* L. (Lepidoptera). Comp. Physiol, A 160:645–656.

Swihart, C.A. & Swihart, S.L. 1970. Colour selection and learned feeding preferences in the butterfly, *Heliconius charitonius* Linn. Anim. Behav. 18:60–64.

Trimen, R. 1887. South African butterflies: A monograph of the extra-tropical species. Vol 1. Nymphalidae. Trubner & Co, London.

Vogel, S. 1954. Blutenbiologische typen als elemente der sippengliederung. Bot. Stud. 1:1–338.

Wiklund, C., Eriksson, T. & Lundberg, H. 1979. The wood white butterfly, *Leptidea sinapis*, and its nectar plants: a case of mutualism or parasitism? Oikos 33:358–362.

PART SIX

Seed dispersal

CHAPTER 14

Modes of dispersal of seeds in the Cape fynbos

EUGENE J. MOLL[1] and BRUCE MCKENZIE[2]
[1]*Department of Management Studies, University of Queensland, Gatton College, Lawes Qld 4343, Australia*
[2]*Botany Department, University of Western Cape, Private Bag X 17, Bellville 7530, South Africa*

Key words: Afromontane forest, Cape Floral Kingdom, heathlands, Mediterranean shrublands

Abstract. An analysis of phytosociological data from the various shrubland communities in the Mediterranean climate zone of South Africa, based on the contribution of species to total canopy cover, shows important differences between community-wide modes of dispersal in different vegetation formations. In undisturbed Renosterveld 39% of important species are dispersed by animals, 23% by wind and 34% passively. Once Renosterveld is disturbed wind-dispersed species increase to 32% at the expense of animal- and passively-dispersed plants. In Strandveld animal-dispersed plants are most important (44%), followed by wind (31%)- and passively (19%)-dispersed plants. On the arid fringes, however, plants dispersed by animals, especially birds, decrease and passively-dispersed succulents increase. In the Succulent Karoo 52% of common species are passively-dispersed and 33% wind-dispersed, with animal-dispersed species comprising only 7%. The heathlands of the Cape are very difficult to analyse because of the high degree of diversity and turn-over rates, thus local dominance patterns are greatly influenced by disturbance history. It seems, however, that passively-dispersed plants comprise some 50% and more in the heathlands, and those dispersed by animals (mainly ants) may locally comprise up to 30% of dominants. Forest species on the other hand are essentially bird-dispersed.

Nomenclature: follows Bond and Goldblatt (1984).

Introduction

The Cape Fynbos region is a biogeographically complex zone containing elements of two floral kingdoms (Cape and Palaeotropic: Good, 1974), five phytochoria (Capensis, Karoo-Namib, Afromontane, Tongaland-Pondoland and Kalahari-Highveld: Werger, 1978), five biomes (Fynbos, Succulent Karoo, Forest, Savanna and Grassland: Rutherford and Westfall, 1986) and five major vegetation formations (Cape fynbos, Renoster shrubland, Karroid shrublands, Forest, Thicket, and Grassland and grassy shrubland: Campbell, 1985).

The climate of the region ranges from a true Mediterranean climate in the west, through an all-year rainfall zone along the south coast, to a summer rainfall zone in the east. Mean annual rainfall varies considerably from the coast inland, and with altitude (from over 3000 mm to less than 200 mm: Schulze, 1972; Fuggle, 1981). The complex geology of the region has led to a small scale mosaic of soils and drainage patterns, and superimposed on this is a palaeoecological history and pattern of peopling of the region that has complicated matters further (Deacon *et al.*, 1983).

It is not, therefore, surprising that there is considerable confusion in the literature concerning the definition of fynbos, and that fynbos has become synonymous with the Cape mediterranean vegetation. Prior to the 1970s very little of this complexity was appreciated both locally or internationally, though Acocks (1953) did note that it was 'a complex vegetation, and to divide it simply into two types was like dividing tropical vegetation into grassland and woodland' (when

M. Arianoutsou and R.H. Groves, Plant-Animal Interactions in Mediterranean-Type Ecosystems, 151–157, 1994.

Acocks had already recognised 68 veld types and 70 major variations in the remainder of the Republic of South Africa!).

With the establishment of the Fynbos Biome Research Programme (Huntley, 1992) our biological knowledge of the region exploded (see, for example, Cowling, 1992), and today we recognise two major vegetation associations in the region. These are the heathland and non-heathland communities, determined primarily by soil nutrient status and climate respectively (Moll, 1991). Coincidently Cape heathland communities occupy a major portion of the Mediterranean climate zone of southern Africa, and have thus been incorrectly interpreted as Mediterranean-type shrublands. In this chapter, the modes of persistence and dispersal of seeds will be analysed for all community types that occur in the Mediterranean climate zone, and the major differences between the heathland and non-heathland communities will be identified.

Major vegetation formation of the Cape Mediterranean region (after Moll et al., *1984)*

Mediterranean shrublands

Renosterveld. According to Moll and Bossi (1983, 1984) some 12% of the region was once Renosterveld. It has been suggested that Renosterveld was probably dominated by grasses in pre-historical times. Today Renosterveld is mostly dominated by a single species, *Elytropappus rhinocerotis*, a secondary, wind dispersed shrub. However, in some localised areas, such as Signal Hill (Joubert, 1991), the vegetation structure and floristic composition is more similar to what it was in the distant past, a number of species (shrubs and grasses) dominating. Signal Hill Renosterveld communities have a total species list in excess of 460 vascular plants and grasses, and *Rhus* spp. form the dominant cover. Because Renosterveld occupies the most nutrient rich soils having a reliable rainfall (granite and/or shale derived), a major problem concerning our understanding of Renosterveld is that only some 7.5% of what once occurred remains today, and these relict patches are all on steep, non-arable land, atypical of what once comprised the Renosterveld landscape.

Strandveld. This occupies the more calcium rich recent sands along the coastal forelands, and once covered some 2% of the region. Strandveld is a complex mosaic and mixture of different growth forms and species that vary considerably from the more arid north to the wetter south. Because the soils are poor and subject to wind erosion only about half the area has been cleared for agriculture.

Succulent Karoo. These dwarf open shrublands occupy some 65% of the region, occurring on the arid fringe. Because of the aridity very little Succulent Karoo has been cleared for cultivation. It is believed, however, that there has been a fairly considerable change in floristic composition during historical times because domestic stock have selectively grazed out the grass component leaving the less palatable shrub component to dominate the landscape (Hilton-Taylor, in prep.). Thus the dominance of succulent shrubs in the region may be an artifact of past grazing regimes (Acocks, 1953).

Non-mediterranean shrublands and Afromontane forest

Heathlands. Some 20% of the area is occupied by Cape heathland vegetation. These shrublands dominate on the oligotrophic and/or dystrophic soils, usually quartzites, laterites or calcareous soils. This is the most complex vegetation type in the Cape with dominance varying considerably according to site characteristics, post-fire age and management history. The plant diversity and degree of endemism of the vegetation is well known. Cowling *et al.* (1992) summarized the major trends and differences in the Cape, and between the Cape and other mediterranean regions. Very little has been cleared for agriculture because the soils are shallow and nutritionally poor, but all areas have been subjected to fires of human origin. Therefore, it is difficult, if not impossible, to reconstruct a picture of what the structure and floristic composition of this heathland component may once have been.

Forest. Less than 1% of the region is covered by forest which occurs today only in higher rainfall areas that are partly protected from fire. In accessible areas all good timber trees were felled in historic times, and there is evidence that the multi-stemmed nature of many individual trees may be a result of disturbance (cutting, fire and wind damage). Recent studies indicate that these Afromontane forest patches are expanding and that without fire they would be more extensive (Luger & Moll, 1993).

Important categories of dispersal modes in the Cape

Ridley (1930) in his seminal work on the dispersal of plants throughout the world has broad categories of dispersal modes, namely wind, water, animals and mechanical means. Within each major category Ridley has numerous sub-categories, but deliberately avoids complex terminology. On the other hand, Van der Pijl (1982) uses much more specific and complex terminology to describe dispersal types. In this review we will look at different Cape mediterranean communities, in some cases at different post-fire successional stages or under one or other management regimes, and discuss the dispersal modes of the major constituents of these communities, ignoring in some cases the perhaps more interesting and specialised but ecologically insignificant anomalies. In this way we demonstrate the marked differences that occur in dispersal modes between major vegetation types, and within the same type because of intrinsic (post-fire successional stages) and extrinsic (management regime) differences.

But first, a listing of the most important dispersal modes to be found in the Cape:

Wind – many plants have tiny seeds, many of which have wings, hairs or other structures that facilitate being carried through the air or rolled along the ground (e.g. Proteaceae, Restionaceae, Asteraceae).

Animals – bird-distributed seeds via fleshy fruits are common in some communities (Forest and Strandveld), as are spiny fruits (Strandveld) that are dispersed on the feet of soft-footed mammals (Moll, 1992). Another interesting mode in the SW Cape is mole-rat dispersal of plants with bulbs (Lovegrove & Jarvis, 1986) and corms, which in some localized communities comprise the bulk of species present (e.g. Sandveld: Mason, 1972). Finally, there are many myrmecochorous plants in the heathlands (e.g. *Leucospermum*, *Phylica*, *Aspalathus*).

Passive – many plants, mainly with small seeds are apparently passively dispersed (e.g. Ericaceae, Orchidaceae, Aizoaceae). In some cases we suspect that rain water wash may further aid distance dispersal.

Regarding persistence of seed banks, this is of vital importance in fire-prone and arid environments such as in the Cape, for both the heathlands and non-heath mediterranean shrublands, because most recruitment occurs post-fire or after a significant rainfall event. Many seeds are stored in the soil either passively or actively (myrmecochorous), and some heathland plants are serotinous, with these serotinous species being the dominant overstorey shrubs in may fynbos (heathland) communities. Serotiny is also a characteristic of the Aizoaceae, the dominant succulent family in the Succulent Karoo. There is little evidence of long-range dispersal mechanisms in most Cape communities (except Forest and to some extent Strandveld: bird dispersal); in fact, most undisturbed vegetation types seem to be characterized by short-distance dispersal mechanisms and a high degree of local endemism.

An analysis of dispersal syndromes

Mediterranean shrublands

Renosterveld. Based on an analysis of the phytosociological data from Boucher (1987, Table 20: 133 relevés), out of some 500 perennially identifiable species, 40% are animal and 15% are wind dispersed. But when species with a Braun-Blanquet cover value of 2 or more (5–25%) are considered, of the 94 spp. 42% are animal, 32% are wind and 22% are passively-dispersed. Of the most widespread species both *Eriocephalus africanus* and *Elytropappus rhinocerotis* are wind dispersed, and *Ischyrolepis capensis* and *Anthospermum aethiopicum* are passively dispersed. Data from Joubert's (1991) Signal Hill survey gives similar general statistics. Of the 66 species (Table 5.2: 54 relevés) 39% are animal, 23% wind and 34% passively-dispersed. An analysis of the 56 species with a Braun-Blanquet cover value of 1 or more (>1%) shows that 52% are animal dispersed species of which, 62% are bird dispersed (32% of the 56 spp.). These statistics are different from the general trend (Boucher, 1987) because Signal Hill has a long history of protection from fire and grazing by domestic stock (±35 years), and bird dispersed species, especially *Rhus lucida*, have become dominant over the last 20 years or so, though of the seven most widespread and common species, five are wind dispersed. On Signal Hill, however, *Elytropappus* is not one of these as this species is indicative of frequently disturbed sites; it is, therefore, an unimportant constituent of the original vegetation.

Strandveld. Based on an analysis of the phytosociological data from Boucher (1987, Table 15: 100 relevés), out of some 300 perennially identifiable species 37% are animal-, 26% wind-dispersed and 29% passively-dispersed. Of the 57 species that have a cover per relevé of more than 5%, 44% are animal-, 31% wind-and 19% passively dispersed. Five of the six most widespread and common species (*Euclea racemosa*, *Rhus laevigata*, *Putterlickia pyracantha*, *Chrysanthemoides incana* and *Rhus glauca*) are all bird-dispersed.

As one moves north along the west coast the climate becomes more arid and the number and importance of bird dispersed species decreases, whilst the succulent element increases (being mainly passively or wind-dispersed). According to Liengme (1988), this increase in the number and cover of succulents is an artifact of grazing by domestic stock. She argues that in pre-historic times when nomadic herders roamed the region there may well have been a little more grass, and certainly more bird-dispersed shrubs in the north (that have been removed over time to provide fuel for cooking, as they are the only major source of woody material in these arid regions).

Succulent Karoo. No extensive phytosociological studies have been done in the Succulent Karoo. Data are available from two local studies, one by Lloyd (1989) in the more arid north where her data from 178 relevés 4 × 4m in size gave 46 perennially identifiable species. Of these, only 7% are animal-, 33% wind- and 52% passively dispersed. In most communities recognized by Lloyd the dominant species were passively-dispersed succulents, but on the more loamy, Nama Karoo sites grasses with wind dispersed seeds were co-dominant with succulents. The other survey was by Smitheman and Perry (1990) from near Worcester in the southwest. They recorded 52 perennial species that were common in 73 relevés, each 5 × 5 m in size. Of these species 15% were animal-, 23% wind- and 48% passively dispersed. These data support personal observations that most of the dominant species in the Succulent Karoo are passively dispersed (perhaps 50%), with wind-dispersed shrubs and grasses comprising some 25%. Bird dispersed shrubs are inconspicuous in the landscape, although in localised areas they may contribute >1% cover.

When one considers the annual and geophytic species which are not perennially identifiable, and for which there are no quantitative data at present, the field impression is that most of the annuals have wind dispersed seeds, and the geophytes have both wind and passively dispersed seeds.

Non-Mediterranean shrubland and forest

Heathlands. The situation with regard to the mode of dispersal of the dominant species in the heathlands is a very complex one to analyse. There are numerous phytosociological studies available, but because of the historical impact of various management regimes, complex local environmental factors, and the various post-fire ages of the various communities, it is not easy to extract satisfactory quantitative data that gives an overview that can be considered average. To illustrate these difficulties we give three examples. The first relates to a shrubland dominated by *Leucadendron laureolum* that was subjected to a fire of high intensity. The 20 year old pre-burn community was dominated by a serotinous shrubby overstorey with post-fire wind-dispersed seeds, and a predominantly passively and/or wind-dispersed restionaceous-ericaceous understorey. Fire killed all re-seeding species and most of the seed bank. The gaps between the few resprouting plants were rapidly covered by a species of *Aspalathus* (ant-dispersed) and *Helichrysum vestitum* (wind-dispersed). The community 3–7 years after the fire was dominated by these two species with *H. vestitum* comprising about 90% of the cover and biomass by the seventh year. Now, some 10 years later, *H. vestitum* has died and the structure and composition of the community is beginning to resemble the pre-burn state except that the canopy cover of *L. laureolum* is only some 15%, not 90% of the area. The second example concerns a ±45 year old *Protea laurifolia* community at Bainskloof which was burnt in a summer wild fire in 1982 (the optimal time of the year for management burns). Prior to the fire *P. laurifolia* shrubs comprised some 80% of the overstorey canopy with an understorey of restios and ericoid shrubs. *P. laurifolia* is wind-dispersed and the understorey species were mainly passively or ant-dispersed. An adjacent fire break, which until 1978 had been burned on a six year cycle, had no obligatory seed-regenerating *P. laurifolia* individuals in it and was dominated by Restionaceae; all individuals resprouted after the fire and mostly had ant- or passively-dispersed seeds. After the 1982 fire which burnt the whole area, scattered *P. laurifolia* individuals established themselves. A second summer wild fire razed the area in 1990. Currently, there are no seedlings of *P. laurifolia* on the sites where it was dominant before 1982 while the fire break has enough seedlings to develop a canopy covering 50% of the site in 10–15 years time. There has, therefore, been a complete elimination of *P. laurifolia* from one site and colonisation of another adjacent site, resulting in a shifting pattern of dominance in the landscape. The third example is also from Bainskloof, where in 1978 a wild fire in summer burnt a fynbos mountainside that was covered by a closed canopy of a ±40 year old *Erica hispidula* community. *E. hispidula* has a tiny seed that is probably passively dispersed, and also may be blown and washed for short distances. This *E. hispidula* community has not re-established itself and we assume that all the tiny seeds were killed in the hot fire. Instead, a mixed fynbos community with restioid and some ericoid plants has formed. Many of these are post-fire resprouting species that comprised an inconspicuous understorey in the pre-fire *E. hispidula* dominated community. The re-seeders present today are a mixture of ant-, passive- and wind-dispersed species.

The only pre- and post-fire data we are aware of, with respect to regeneration modes of the post-fire vegetation, are those of de Lange (1992). She analysed post-fire data from 50 relevés, 5 × 10 m in size. These relevés were from nine fynbos communities (and they had been surveyed before the fire too). Of the 767 species recorded, some 1% were monocarpic re-seeders. In the post-fire relevés, 64–98% of the species were the same as those that were present before the fire, representing a slight floristic change. If the area outside the relevé was searched for its flora all the "missing" species would have been encountered in the neighbourhood. Some 51–67% of species resprouted, a much lower figure than that quoted by Kruger (1987), whose study sites were all in mountain catchment areas burned regularly and managed by the Department of Forestry.

An analysis of the 177 species which contributed more than 1% to the cover of the relevés in de Lange's (1992) pre-fire survey indicates that 27% are animal- (mainly ant), 14% are wind- and 55% are passively-dispersed. This contrasts with the work of McDonald (1983) who identified 21 communities from 170 relevés in an area of approximately equivalent size and post-fire age. From an analysis of the 241 species recorded by McDonald, some 8% are animal-, 21% wind- and 61% passively dispersed. From these figures and other similar observations we claim that the percentage contribution of each dispersal mode changes, but, in general, there is always a good proportion of dominant species that are ant- and wind-dispersed. The fynbos landscape, however, is mostly comprised of passively-dispersed species, though locally, wind-dispersed species (e.g. serotinous Proteaceae) can be dominant. These average statistics are influenced by the extent to which resprouters dominate in local communities. In some areas, there is virtually little or no post-fire seed regeneration, whilst in other areas there may be considerable regeneration from seed.

Forest

All species of canopy trees are bird-dispersed. A notable exception is *Cunonia capensis*, which is wind-dispersed. Trees and shrubs of the understorey are also bird-dispersed.

Conclusions

We have indicated the important differences in proportion of dispersal modes, not only between the various vegetation types in the Mediterranean climate zone, but also as a consequence of various intrinsic and extrinsic factors. The changes that occur as a result of these factors, especially the proportion of seeds available for dispersal by animals (primarily birds and ants) are obviously important to the understanding of dispersion patterns of animals. An in-depth study of plant-animal relationships is required if we are to enhance our understanding of the temporal and spatial dynamics within the vegetation types of the Mediterranean climate zone of South Africa.

References

Acocks, J.P.H. 1953. Veld types of South Africa. Mem. Bot. Surv. S. Afr. 28:1–192.

Bond, P. & Goldblatt, P. 1984. Plants of the Cape Flora: a Descriptive Catalogue. Supplement to J. S. Afr. Bot. 13:1–455.

Boucher, C. 1987. A phytosociological study of transects through the Western Cape coastal foreland, South Africa. Ph.D. Thesis, University of Stellenbosch, South Africa.

Campbell, B.M. 1985. Montane vegetation structure in the fynbos biome: structural classification and adaptive significance of structural characters. Ph.D. Thesis, University of Utrecht, The Netherlands.

Cowling, R.M. (ed.) 1992. The Ecology of Fynbos: Nutrients, Fire and Diversity. Oxford University Press.

Cowling, R.M., Holmes, P.M. & Rebelo, A.G. 1992. Plant diversity and endemism. In: Cowling, R.M. (ed.). The Ecology of Fynbos: Nutrients, Fire and Diversity. Oxford University Press, pp. 63–112.

De Lange, C. 1992. Floristic analysis of the Vogelgat Nature Reserve, Cape Province, South Africa. M.Sc. Thesis, Botany Dept., University of Cape Town, South Africa.

Deacon, H.J., Hendey, Q.B. & Lambrechts, J.J. (eds.). 1983. Fynbos palaeoecology: a preliminary synthesis. S. Afr. Nat. Sci. Prog. Report 75:1–216.

Fuggle, R.F. 1981. Macro-climatic patterns within the Fynbos Biome. Final Report Nat. Prog. Environ. Sci. 1–116.

Good, R. 1974. The geography of flowering plants. 4th ed. Longman, London.

Hilton-Taylor, C. (in prep.) Phytogeography of the winter rainfall karoo. Ph.D. Thesis, University of Cape Town, South Africa.

Huntley, B.J. 1992. The Fynbos Biome Project. In: Cowling, R.M. (ed.). The Ecology of Fynbos: Nutrients, Fire and Diversity. Oxford University Press, pp. 1–5.

Joubert, C. 1991. History and description of contemporary vegetation, Signal Hill, Cape Town. M.Sc. Thesis, University of Cape Town, South Africa.

Kruger, F.J. 1987. Succession after fire in selected fynbos communities of the southwestern Cape. Ph.D. Thesis, University of Witwatersrand, Johannesburg, South Africa.

Liengme, C.A. 1988. Pastures, past and present: utilization of West Coast Strandveld. M.Sc. Thesis, University of Cape Town, South Africa.

Lloyd, J.W. 1989. Phytosociology of the Vaalputs radioactive waste disposal site, Bushmanland, South Africa. S. Afr. J. Bot. 55:372–382.

Lovegrove, B.G. & Jarvis, J.U.M. 1986. Coevolution between mole-rats (Bathyergidae) and a geophyte, *Micranthus* (Iridaceae). Cimbebasia (A) 8:79–85.

Luger, A.D. & Moll, E.J. 1993. Fire protection and Afromontane forest expansion in Cape fynbos. Biol. Conserv. 64:51–56.

Mason, H. 1972. Western cape Sandveld flowers. C. Struik (Pty) Ltd., Cape Town.

McDonald, D.J. 1983. The vegetation of Swartboschkloof, Jonkershoek, Cape Province, South Africa. M.Sc. Thesis, University of Cape Town, South Africa.

Moll, E.J. 1991. Mediterranean vegetation in the Cape Province, South Africa: a review of recent concepts. Ecol. Medit. 16:291–296.

Moll, E.J. 1992. Convergence of podochorous diaspores in the SW Cape – a possible connection with hominids? In: Thanos, C.A. (ed.). Plant-Animal Interactions in Mediterranean Type Ecosystems. Proc. VI International Conference on Mediterranean Climate Ecosystems, Athens, Greece, pp. 263–268.

Moll, E.J. & Bossi, L. 1983. 1:1 000 000 Vegetation map of the Fynbos Biome. Chief Director Surveys and Mapping, Mowbray, South Africa.

Moll, E.J. & Bossi, L. 1984. Assessment of the extent of the natural vegetation of the Fynbos Biome of South Africa. S. Afr. J. Sci. 80:351–352.

Moll, E.J. Campbell, B.M., Cowling, R.M., Bossi, L., Jarman, M.L. & Boucher, C. 1984. A description of major vegetation categories in and adjacent to the fynbos biome. S. Afr. Nat. Sci. Prog. Report 83: 1–28.

Ridley, H.N. 1930. The dispersal of plants throughout the world. Reeve & Co. Ashford, Kent.

Rutherford, M.C. & Westfall, R.H. 1986. Biomes of southern Africa: an objective categorization. Mem. Bot. Surv. S. Afr. 54: 1–98.

Shulze, B.R. 1972. South Africa. In: Griffiths, J.F. (ed.). World Survey of Climatology. Vol. 10. Climates of Africa. Elsevier, Amsterdam, pp. 501–586.

Smitheman, J. & Perry, P. 1990. A vegetation survey of the Karoo National Botanic Garden Reserve, Worcester. S. Afr. J. Bot. 56:525–541.

Van der Pijl, L. 1982. Principles of dispersal in higher plants. Springer-Verlag, Berlin.

Werger, M.J.A. 1978. Biogeographical division of southern Africa. In: Werger, M.J.A. (ed.). Biogeography and Ecology of Southern Africa. Dr W. Junk, The Hague, pp. 145–170.

CHAPTER 15

Why are there so many myrmecochorous species in the Cape fynbos?

R. M. COWLING,[1] S.M. PIERCE,[2] W.D. STOCK[1] and M. COCKS[1]
[1]*Department of Botany and* [2]*Bolus Herbarium, University of Cape Town, Rondebosch* 7700, *South Africa*

Key words: ant, dispersal, diversification

Abstract. The exceptionally high incidence of myrmecochory (ant-dispersed species) in Cape fynbos and similarly fire-prone vegetation on nutrient-poor soils in Australia has aroused much interest. An ecological advantage of myrmecochory on both continents is the removal of seeds to sites safe from seed predation. By failing to address evolutionary questions, ecological hypotheses do not explain the very high numbers of myrmecochorous species in fynbos. We show that myrmecochores are seldom dominant in fynbos communities although species numbers are invariably high. We also show that myrmecochores are not a random assemblage with regard to biological traits. Most myrmecochores are dwarf to low shrubs, obligately reseeding from relatively large seeds. Limited data suggest that myrmecochores have small, transient seed banks and are vulnerable to fire-induced local extinction. Myrmecochores were also significantly over-represented amongst a lowland neoendemic flora, suggesting that lineages possessing this trait are associated with recent diversification. Finally, in a general survey of the fynbos flora, we found that myrmecochorous genera were significantly more speciose than genera with other dispersal modes.

We conclude that the major ecological advantage of myrmecochory is the removal of large, precious seeds to sites safe from predators. Large seed size ensures seedling establishment in the nutrient-poor summer dry fynbos environment. Myrmecochores produce fewer seeds, however, and have smaller, less persistent seed banks than species with other dispersal modes. These traits, in combination with obligate reseeding and short dispersal distances, result in myrmecochore lineages being vulnerable to fire-induced population reduction and fragmentation. These processes also, and incidentally, promote diversification.

Nomenclature: follows Bond & Goldblatt (1984).

Introduction

Myrmecochory, the dispersal of seeds by ants, is especially well represented in vegetation on infertile soils in the Mediterranean-climate regions of Australia and the Cape, South Africa (Berg, 1975; Slingsby & Bond, 1981; Milewski & Bond, 1982; Bond & Slingsby, 1983). In the Cape, it is estimated, largely on the basis of seed morphology, that myrmecochory is present in 29 families, 78 genera and about 2500 species, i.e. about 30% of the fynbos flora (Bond & Slingsby, 1983; Breytenbach, 1988; Bond *et al.*, 1991). A typical regional flora from fynbos may have upwards of 25% ant-dispersed species (Fig 15.1), a feature shared only with analogous vegetation in Australia (Milewski & Bond, 1982; Willson *et al.*, 1990; Westoby *et al.*, 1991).

Myrmecochory in South Africa

In the Cape, myrmecochores are concentrated in fynbos, a sclerophyllous, fire-prone shrubland confined to nutrient-poor soils. Myrmecochory is largely absent from vegetation on nutrient-rich soils in the Cape Mediterranean-climate region. A few ant-dispersed species may occur in renoster shrubland (Le Maitre & Midgley, 1992; R. M. Cowling, pers. obs.), a fire-prone vegetation on base-rich soils dominated by small-leaved shrubs (largely Asteraceae) with wind-dispersed seeds and which has an understorey of Poaceae and geophytes (Boucher & Moll, 1981; Campbell, 1985). Myrmecochores are extremely rare in the closed forest and thicket vegetation of tropical affinity (Milewski & Bond, 1982; Le Maitre & Midgley, 1992) which penetrates into the Cape

M. Arianoutsou and R.H. Groves, Plant-Animal Interactions in Mediterranean-Type Ecosystems, 159–168, 1994.

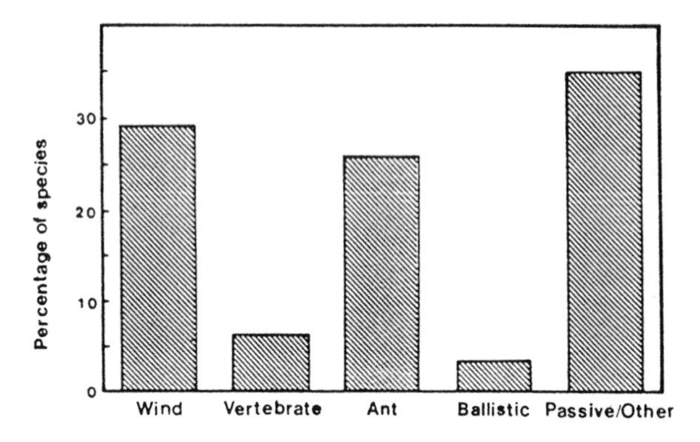

Fig. 15.1. Percentage of plant species ($n = 562$) in different dispersal categories. Data are from species lists compiled from 12 0.1 ha plots sampled in fynbos on the Agulhas Plain (for details see Cowling, 1990; Cowling & Holmes, 1992).

on fertile and fire-protected sites (Cowling & Holmes, 1992). These communities are dominated by large-leaved trees and shrubs, mostly members of the Anacardiaceae, Celastraceae, Ebenaceae, and Oleaceae, with vertebrate-dispersed seeds (Phillips, 1931; Milewski & Bond, 1982; Cowling, 1984; Le Maitre & Midgley, 1992). The few vertebrate-dispersed species occurring in fynbos (see Le Maitre & Midgley, 1992; Fig 15.1) are mostly forest and thicket elements which require the prolonged absence of fire for successful establishment (Manders & Richardson, 1991). The semi-arid karroid shrublands which occur in drier and more fertile sites than fynbos also have low numbers of myrmecochores. Dispersal spectra are dominated by wind-dispersed species (Asteraceae) and those with hygrochastic capsules (Mesembryanthemaceae) (Hoffman & Cowling, 1987; Hartmann, 1991; Le Maitre & Midgley, 1992). Genera such as *Zygophyllum* (Zygophyllaceae) and *Osteospermum* (Asteraceae) which straddle the fynbos-karoo boundary, have ant-dispersed species in the former and wind-dispersed species in the latter (Bond & Slingsby, 1983). The situation in the Cape is quite different from Australia where myrmecochores are relatively common in woodlands on fertile sites (Westoby *et al.*, 1990) and in arid shrublands (Davidson & Morton, 1981).

Current hypotheses explaining concentrations of myrmecochores

There has been much recent research on the ecological importance of myrmecochory for plants in the Cape and Australia (reviewed by Bond *et al.*, 1991 and by Westoby *et al.*, 1991, respectively). Many hypotheses on the benefits of dispersal by ants have been rejected on both continents. There is experimental evidence, however, that ants remove seeds to sites safe from small mammal predators in the Cape (Bond & Breytenbach, 1985) and similarly from ant predators in Australia (Westoby *et al.*, 1991).

A question which has received much attention on both continents is why myrmecochores are much more frequent on infertile than fertile soils (Milewski & Bond, 1982; Westoby *et al.*, 1990). Are there differences in dispersal rate and predation pressure between the two soil types? There is some experimental evidence refuting these hypotheses (Breytenbach, 1988; Mossop, 1989) but more data are required on comparative predation pressures. Westoby *et al.* (1990) argued that the high frequency of ant-dispersed species on infertile soils in Australia may be a secondary correlate of plant stature and seed size. In other words, low-nutrient soils may favour growth forms and seed sizes which, in turn, favour dispersal by ants. Their study showed that species with lower (<2 m) stature and, to a lesser extent, smaller seed size were more likely to be adapted for dispersal by ants than by vertebrates. They found that on fertile soils, myrmecochorous shrubs are replaced by vertebrate-dispersed trees. A similar argument related to form has been developed for fynbos by Pierce (1990): since nutrient-poor soils select for small leaves (Cowling & Campbell, 1980), and since small leaves are borne on slender shoots (Midgley & Bond, 1989), vertebrate dispersal is likely to be rare since this syndrome requires both stout twigs and large (reward-rich) fruits. These hypotheses therefore suggest that because other dispersal modes are not feasible, dispersal by ants is common in nutrient-poor environments (see also Bond *et al.*, 1991).

The association between low plant stature and myrmecochory is interesting. Westoby *et al.* (1990) suggest that ant dispersal is a 'cheap but adequate' mode for low plants. Since short dispersal distances are adequate for low plants, there is no selection for large and more costly seeds suitable for vertebrate-dispersal. There is no basis to the contention that ant dispersal is less costly than bird dispersal, however. Preliminary data for closely related fynbos species in the Asteraceae indicate that the carbon, nitrogen and phosphorus costs of vertebrate-dispersal are less than in the case of ant-dispersal and both modes are costlier than wind-dispersal (Stock & Bond, 1992). There are other problems with the plant-size hypothesis. Firstly it fails to explain the predominance of low plants (<1 m) in renoster shrubland with mainly wind dispersal (Le Maitre & Midgley, 1992). Secondly, in fynbos there are many large-leaved, bird-pollinated shrubs in the Proteaceae which, contrary to the predictions of the small-size hypotheses, lack vertebrate-dispersed seed. Finally, and more generally, there are no convincing arguments regarding the benefits of long or short distances for seed dispersal in the fire-prone environments where myrmecochores are concentrated (but see Andersen, 1988; Yeaton & Bond, 1990).

Current hypotheses do not answer the question: why are there so many myrmecochores in fynbos? This is because explanations to date have failed to consider evolutionary as well as ecological processes. In this chapter we consider the relationships between seed biology, fire, population dynamics, population genetic processes and diversification rates of fynbos lineages in order to provide an evolutionary or ultimate explanation for the high number of myrmecochores in fynbos communities and landscapes. We invoke Vrba's (1980) effect hypothesis which states that 'selection for proximal fitness of organisms may also and incidentally drive speciation rates'. In other words, we suggest that the biological traits associated with the myrmecochorous syndrome in fynbos, which confer selective advantage to individuals, incidentally results in rapid diversification and the accumulation of ant-dispersed taxa in fynbos landscapes.

In developing our argument we ask several questions:

1) What is the relationship between cover abundance (ecological success) and species numbers (evolutionary success) of myrmecochores in fynbos communities?
2) What are the growth form, seed biological and fire resistance correlates of myrmecochory in fynbos?
3) Are myrmecochores over-represented among fynbos neoendemics?
4) Are ant-dispersed genera more speciose than those with other dispersal modes?

Some of these questions have been addressed for Australian vegetation (Rice & Westoby, 1981; Willson *et al.*, 1990; Westoby *et al.*, 1990) but the synthesis we present is new. Although our data base has deficiencies, the analysis does suggest new approaches for exploring the debate regarding the high incidence of ant dispersal in southern hemisphere vegetation on infertile compared with fertile soils.

Table 15.1. The mean ± S.D. (range) relative cover, number and proportion of myrmecochorous species in 12 0.1 ha plots located in fynbos communities encompassing a wide range of environments in the Mediterranean-climate region of the southwestern Cape, South Africa. Rainfall range for plots: 350–1200 mm yr^{-1}, altitude: 5–960 m, geology: quartzite, siliceous sand, laterite, granite, limestone and calcareous sand. Data from Kruger (1979) and R.M. Cowling (unpublished)

	Cover (%)	No. spp.	Proportion of total spp. (%)
Myrmecochorous species			
Woody spp. only	11.8 ± 14.6 (0.3–43.2)	7.9 ± 4.5 (1–16)	22.0 ± 8.7 (5.5–37.5)
Woody and non-woody spp.	18.4 ± 13.1 (1.3–50.0)	13.9 ± 5.1 (6–20)	24.0 ± 9.1 (12.5–37.1)

What is the relationship between cover abundance and species number of myrmecochores in fynbos?

We answered this question by assessing the incidence of myrmecochory in 12 0.1 ha plots sampled in fynbos encompassing a wide range of habitats in the southwestern Cape (Table 15.1). The presence of ant dispersal was determined from a list of genera published by Slingsby and Bond (1981) and Bond and Slingsby (1983), discussion with colleagues (W.J. Bond & H.P. Linder) and personal observations. These determinations are limited since they are based largely on seed morphology (the presence of an elaiosome): direct observations of ant dispersal in fynbos are confined to a few species of the various genera and families. Furthermore, we classified as ant-dispersed all the species of certain genera, (mostly graminoids – *Ficinia*, *Tetraria* in the Cyperaceae; *Cannomois*, *Ceratocaryum*, *Hypodiscus*, *Mastersiella*, *Willdenowia* in the Restionaceae), which include both myrmecochorous and non-myrmecochorous species. Thus our data probably overestimate the incidence of ant dispersal.

For woody species the proportion of myrmecochores in a plot exceeded by two-fold their relative cover (Table 15.1). This value is greatly reduced when herbaceous myrmecochores are included. These data are broadly similar to those recorded for Australian sclerophyll vegetation (Rice & Westoby, 1981).

Our data suggest that for woody plants at least, myrmecochores are not an ecologically dominant component in many fynbos communities although species numbers are proportionally high. Indeed, the woody component of most fynbos communities is dominated either by tall, serotinous shrubs with wind-dispersed seeds or ericaceous shrubs with dust-like seeds (Le Maitre & Midgley, 1992). Herbaceous myrmecochores comprising mainly Cyperaceae and Restionaceae may be ecologically dominant in certain sites (see Campbell, 1985). However, and of relevance to our subsequent arguments, most woody myrmecochores are obligate reseeders after fire, whereas herbaceous species mostly resprout (Cowling & Holmes 1992; Le Maitre & Midgley, 1992).

What are the biological correlates of myrmecochory?

Are fynbos myrmecochores a random assemblage in terms of growth form, seed biological traits and fire resistance mode? We did not have data suitable for a single contingency analysis using all variables (cf. Westoby *et al.*, 1990). Our analyses are restricted to dispersal mode × trait contingency tables and other simple statistical tests.

Growth form

In a flora from a lowland fynbos landscape, myrmecochores were not randomly distributed with respect to growth form (Fig 15.2; $\chi^2 = 77.78$, df = 7, $P < 0.0001$). Ant-dispersed species were over-represented among low (<1.0 m) shrubs and graminoids, equally represented among medium height (1–2 m) shrubs, and under-represented among all other growth forms. The concentration of myrmecochores in low plants, especially shrubs, is consistent with data from sclerophyll vegetation in southeastern Australia (Rice & Westoby, 1981; Westoby *et al.*, 1990). The high number of graminoid myrmecochores (Cyperaceae, Restionaceae) may be unique to fynbos, however, although categorisation into dispersal mode is problematic for these taxa (see above).

Seed size and number

There is a strong selective force to produce large, protein-rich seeds in the fire-prone, nutrient-poor fynbos environment (Stock *et al.*, 1990; Le Maitre & Midgley, 1992). Large, resource-rich seeds enable rapid seedling growth and the avoidance of drought-induced mortality during the dry summer months. For example, Kilian (1991) recorded a two-fold higher seedling mortality during the first summer after establishment of the passively dispersed and small-seeded shrub, *Passerina paleacea*, compared with the larger-seeded and morphologically similar myrmecochore, *Phylica ericoides* (Table 15.2; see also Pierce, 1990). Superimposed on the selection for larger seed size is a constraint on seed number. Nutrient-poor soils are often associated with a high frequency of species with small leaves (Beadle, 1966; Cowling & Campbell, 1980; Givnish, 1987). This results in an allometric constraint (Primack, 1987; Midgley & Bond,1989) on seed size: small leaves are borne on small twigs which bear small seeds (Pierce, 1990). It can be argued that this constraint on the selection for larger seeds has resulted in a proliferation of intermediate-sized seeds (2–25 mg) which are optimal for dispersal by ants (Pierce, 1990). Ant dispersal has evolved to ensure that these precious seeds are buried in sites safe from vertebrate (Bond & Breytenbach, 1985) and ant predators (Westoby *et al.*, 1991; Pierce & Cowling, 1991). Thus, within the small-leaved fynbos shrub guild, we would expect ant-dispersed seed to weigh more than seed dispersed by other means. It should be noted that species

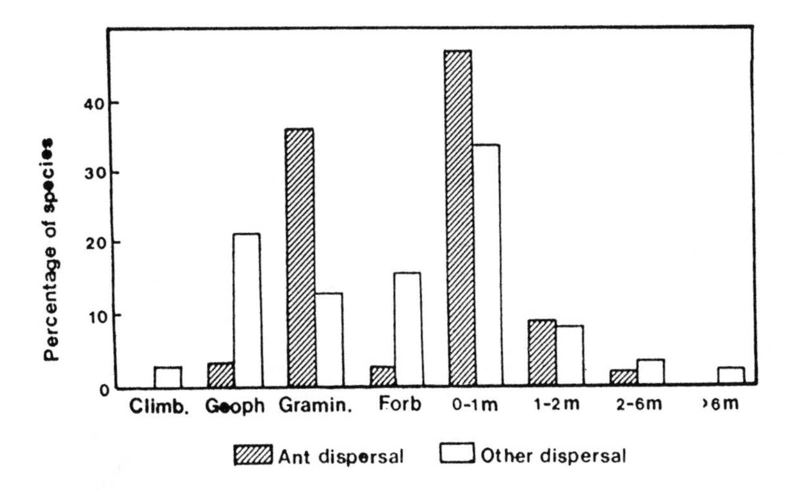

Fig. 15.2. Percentage of plant species ($n = 562$) in different growth form categories dispersed by ants and other means. Height categories (in m) refer to woody plants. Data for Fig. 15.1.

with vertebrate-dispersed seed are almost totally absent from this guild.

We compared the embryo plus endosperm fresh mass (i.e. the 'reserve' (Westoby *et al.*, 1990)) of seeds of 15 ant-dispersed species with 16 species dispersed by wind, ballistic or passive means ($n = 50$ seeds per species) (Fig. 15.3). For several very small-seeded species (<0.25 mg), all non-myrmecochores, it was not feasible to remove the seed coat. The seed resources, particularly nitrogen and phosphorus, which are crucial for seedling establishment in nutrient-poor soils, are concentrated in this reserve (Stock *et al.*, 1990, W.D. Stock & M. Cocks, unpubl.). Our sample includes only small-leaved (leptophyll and smaller) shrubs (0.5–3.0 m) in order to minimise allometrically scaled variation in seed mass. Reserve mass was unrelated to mean shrub height

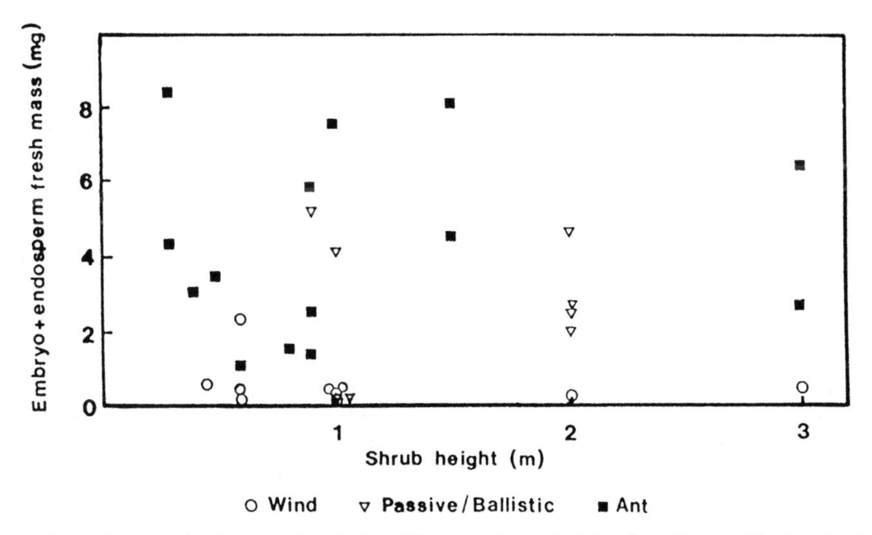

Fig. 15.3. Embryo plus endosperm fresh mass of seeds ($n = 50$) in relation to height of small-leaved fynbos shrub species belonging to different dispersal categories. Wind-dispersed genera (9); passive/ballistic genera (7); ant-dispersed genera (15).

(Spearmans $r = -0.079$, $P = 0.58$) (cf. Thompson & Rabinowitz, 1989). Myrmecochores had a reserve mass which was on average 2.6 times greater than other dispersal modes and this difference was highly significant ($t = 3.30$, $P < 0.005$).

What are the implications for seed numbers? Generally, there is a trade-off between seed size and number: among plants of similar size, those that produce larger seeds have smaller seed crops (Harper, 1977). Analysis of the data available on seed production of small-leaved shrubs (Table 15.2) indicates that fynbos myrmecochores produce larger yet fewer seeds and have smaller seed banks than morphologically similar non-myrmecochores (Pierce & Cowling, 1991). As a result of these small and weakly persistent (Pierce & Cowling, 1991) seed banks, as well as short dispersal distances (Bond *et al.*, 1991), populations of obligatory reseeding myrmecochores are vulnerable to fire-induced population reduction or local extinction. Pierce and Cowling (1991) observed the recruitment failure of an ant-dispersed shrub as a result of the pre-fire extinction of its seed bank. Interestingly, low myrmecochorous shrubs, along with ericaceous shrubs, were identified as a species group highly vulnerable to extinction on fynbos habitat fragments (Cowling & Bond, 1991). Moreover, many small-leaved, ant-dispersed shrub taxa occur in small local populations which are threatened with extinction (Hall & Veldhuis, 1985; Slingsby & Bond, 1985). On the other hand, many wind-dispersed, small-leaved shrubs are fire resilient by migration (Grubb & Hopkins, 1986) and this guild includes most widespread, pioneer fynbos shrubs (e.g. *Elytropappus*, *Metalasia*, *Stoebe*) (Cowling *et al.*, 1986; Pierce & Cowling, 1991). Other fire-resilient non-myrmecochores maintain very large seed banks (e.g. *Passerina*, Table 15.2).

Table 15.2. Seed characteristics of myrmechochorous and non-myrmecochorous small-leaved, fynbos shrubs

	Canopy cover in site (%)	Seed prod. per m^2 canopy	Seed bank (no. m^{-2})	Seed size (mg)
Myrmechochores				
*Agathosma apiculata**	30	5438	61.3	4.6
*A. stenopetala**	7.5	5055	18.7	2.1
*Muraltia squarrosa**	15	202	114	3.7
Phylica ericoides~	15.3	454	278	2.5
P. stipularis +	20	771	<1	24.0
Non-myrmecochores				
*Metalasia muricata**W	18	17508	56	0.6
M. muricata + W	8	2853	15.5	0.6
Passerina palacea~P	19.7	13253	441–737	0.8
*Passerina vulgaris**P	22	44000	629	1.3

* from direct seed counts (Pierce & Cowling, 1991)
~ from direct seed counts (Kilian, 1991)
\+ from seedling counts after fire (Musil, 1991)
W = wind-dispersal; P = passive dispersal

Fire resistance

Data are inadequate to test the hypothesis that obligate reseeding and resprouting are equally frequent among fynbos myrmecochores and non-myrmecochores, i.e. whether myrmecochores are more likely to be reseeders than resprouters. All large ant-dispersed shrub genera (e.g. *Agathosma*, *Cliffortia*, *Diosma*, *Muraltia*, *Phylica*, *Serruria*) have a majority (ca 60–90%) of species killed by fire (Williams, 1982; Le Maitre & Midgley, 1992, A. Bean, pers. comm.; R.M. Cowling, pers. obs). The same is true of most non-myrmecochorous shrubs (e.g. Ericaceae, *Metalasia*, *Passerina*, *Protea*, etc.), however, with the exception of vertebrate-dispersed species, which almost all resprout (Le Maitre & Midgley, 1992). By contrast, most myrmecochorous graminoids are resprouters (pers. obs.) but so are their non-myrmecochorous counterparts (Linder & Ellis, 1990).

Is myrmecochory over-represented among neoendemics?

Since neoendemics are the products of recent diversification reflecting contemporary selective forces (Major, 1988), patterns should identify dispersal modes associated with lineage turnover (Cowling & Holmes, 1992).

In order to answer this question we analyzed the frequency of neoendemics in relation to dispersal categories for a flora from the Agulhas Plain (Table 15.3). This lowland fynbos region has many locally endemic species which are confined to sediments and soils deposited since the mid Pliocene, i.e. they are neoendemics (Cowling

Table 15.3. The association between endemism and dispersal mode of woody species in the Agulhas Plain. ( ) = % of total species (from Cowling & Holmes, 1992). See text for characterization of neondemics

Endemism	Dispersal mode	
	Myrmecochores	Others
Neoendemics	25 (17)	121 (83)
Others	21 (5)	398 (95)

& Holmes, 1992). Ant-dispersed species were significantly over-represented among neoendemics ($\chi^2 = 21.24$, df = 1, $P < 0.0001$). This suggests that myrmecochore lineages have undergone proportionally more diversification in the Quaternary relative to lineages with other dispersal modes. Slingsby and Bond (1985) suggest that the short dispersal distance of myrmecochores might promote geographical isolation and thus enhance diversification rates. We argue later in this chapter that a broader spectrum of biological traits have incidentally promoted diversification of ant-dispersed lineages.

How rich are ant-dispersed genera?

Is the higher diversification of myrmecochores reflected more generally? That is, are there higher numbers of species in myrmecochorous genera versus genera with other dispersal modes? In order to answer this question we analysed the flora of the Cape Floristic Region. We compared the frequency of different categories of richness (i.e. 0–9 spp. per genus, etc.) of 31 genera with ant-dispersed seed and 76 genera with other dispersal mechanisms (Table 15.4). We confined our analysis to small-leaved and predominantly obligatory reseeding shrub genera. Data are from Bond and Slingsby (1983) (myrmecochorous genera) and Bond and Goldblatt (1984) (size of genera).

Myrmecochores were not randomly distributed with respect to categories of genus size (Table 15.4; $\chi = 11.31$, df = 2, $P < 0.005$). The proportion of large (>40 spp.) genera was four times higher for myrmecochores than non-myrmecochores. Large ant-dispersed genera were *Phylica* (133 spp.; Rhamnaceae), *Agathosma* (130 spp.; Rutaceae), *Cliffortia* (106 spp.; Rosaceae), *Muraltia* (106 spp.; Polygalaceae), *Thesium* (79 spp.; Santalaceae) and *Serruria* (48 spp.; Proteaceae). Large non-myrmecochorous genera included *Erica* (528 spp.; Ericaceae), *Aspalathus* (245 spp.; Fabaceae), *Gnidia* (47 spp.; Thymeleaceae), and *Euryops* (46 spp.; Asteraceae).

Table 15.4. The association between dispersal mode and genus in the Cape Floristic Region. The analysis includes only small-leaved and predominantly obligately reseeding genera. [Data from Bond & Slingsby (1983) and Bond & Goldblatt (1984)]. Percentages given in brackets

	No. of species per genus (%)		
	0–9	10–39	>40
Myrmecochores	14 (44)	12 (38)	6 (18)
Other dispersal	64 (74)	18 (21)	4 (5)

Discussion

Most ant-dispersed taxa in fynbos communities and landscapes are low, small-leaved shrubs which comprise a high proportion of the species pool but occur in small and often scattered populations (metapopulations) and occupy a relatively small proportion of the total biomass. A much smaller number of woody myrmecochores are tall, large-leaved shrubs belonging mainly to the genera *Leucadendron*, *Leucospermum*, *Mimetes* and *Paronomus* in the Proteaceae. Almost all research has concentrated on this family (Bond *et al.*, 1991). In terms of herbs, nearly all the ant-dispersed fynbos species are evergreen graminoids which may be ecologically dominant in certain sites. There has been no research on this group and the assessment of their abundance and diversity is problematic. As is the case for Australian sclerophyll vegetation (Rice & Westoby, 1981), myrmecochores seldom dominate fynbos communities.

On fertile sites in the Cape, namely renoster shrublands, myrmecochores are replaced largely by small-seeded, wind-dispersed species. This is in contrast to Australia where myrmecochores are replaced by vertebrate-dispersed species (Westoby *et al.*, 1990). In the Cape, vertebrate-(mainly bird) dispersed species tend to be restricted to forest or thicket on fertile, fire-protected sites. They are nearly all tropical-derived, large-leaved tall shrubs. These communities are not fire-prone (Campbell, 1985 Manders, 1990; Van Wilgen *et al.*, 1990; Manders & Richardson, 1992). We interpret the incidence of frugivory as a requirement for directed dispersal within forest (Knight, 1988) and for colonization of fynbos during extended fire-free intervals (Manders & Richardson 1991; R.M. Cowling, unpubl.).

In the case of renoster shrublands, why are myrmecochores replaced by wind-dispersed species? We believe the answer lies in soil nutrients. In fynbos soils which are relatively nutrient-poor, large and nutrient-rich seeds would confer high seedling survival. The cost of resource-rich seed production, however, is small seed crop size, thereby necessitating adaptations (e.g. myrmecochory) to minimize seed predation. There would be no selection for large seeds on the more fertile renoster shrubland soils: small, wind-dispersed seeds are adequate for post-fire population replacement (Levyns, 1929, 1935).

There remains the problem of explaining the high numbers of ant-dispersed species within fynbos communities. Clearly plants with other dispersal modes, smaller seeds and larger seed crops are ecologically successful in fynbos. Indeed, the higher seedling mortality of these species is offset by higher seedling numbers, resulting in greater per capita population growth after some fires (Pierce, 1990; Kilian, 1991). We suggest that selection for large seed size (and myrmecochory) has implications for diversification rates.

Obligate reseeding plants with short dispersal distances and which produce small and weakly persistent seed banks are vulnerable to fire-related population reduction and extinction (Pierce & Cowling, 1991). Small, widely scattered and genetically isolated populations are prone to catastrophic diversification, especially when isolated in edaphically or climatically marginal sites (Lewis, 1962; Raven, 1964; Stebbins & Major, 1965; Cowling & Holmes, 1992). The fynbos landscape is characterized by extreme edaphic and climatic heterogeneity over short distances (Kruger, 1979; Thwaites & Cowling, 1988) which would promote this type of ecological diversification (Linder, 1985; Linder & Vlok, 1991; Cowling & Holmes, 1992). Thus, in obligate reseeding fynbos lineages, selection for large, ant-dispersed seeds, which may be adaptive at the organism level, incidentally causes the multiplication of taxa. This is a special case of Vrba's (1980) effect hypothesis which states that 'selection for proximal fitness may also, and incidentally, drive speciation'. Traits, such as myrmecochory, may predominate in a flora not because they confer special ecological advantages, but because high diversification rates may overwhelm low survival rates. The generalization is that lineages that are extinction-prone are also prone to diversification (Vrba, 1980). Clades associated with small, well-dispersed seeds and/or large, persistent seed banks produce taxa that are resistant to extinction, and hence are resistant to diversification. The same may be said of sprouting, which, despite its obvious advantage in a fire-prone vegetation (James, 1984), is not common in species-rich fynbos lineages (Le Maitre & Midgley, 1992).

We have argued that the high local and regional richness of fynbos myrmecochores is a result of incidental diversification. Ant dispersal is a highly derived trait in *Leucadendron* (Proteaceae) (Midgley, 1987) and much of the recent diversification in fynbos could be associated with the evolution of myrmecochory. There has been considerable diversification in lineages associated with a different suite of biological traits, however. Striking examples are *Erica* and *Aspalathus*. Both of these genera have microsymbiont-mediated nutrient uptake and there is some evidence for microsymbiont-host specificity (Cowling *et al.*, 1990). These authors have suggested that diversification in these genera could result from selection to acquire new microsymbionts in different soil types. Clearly, many different traits may be associated with diversification in fynbos.

We believe that much of the controversy regarding the importance of myrmecochory in fynbos and similar vegetation in Australia has resulted from an emphasis on the ecological significance of this trait without consideration for evolutionary implications. As pointed out by Fowler and McMahon (1982), the preponderance of species with particular traits in landscapes may be a stronger reflection of incidental diversification than contemporary or historical ecological conditions.

Acknowledgements

H.P. Linder and W.J. Bond commented on an earlier draft.

References

Andersen, A.N. 1988. Dispersal distance as a benefit of myrmecochory. Oecologia 75:507–511.

Beadle, N.C.W. 1966. Soil phosphate and its role in molding segments of the Australian flora and vegetation, with special reference to xeromorphy and sclerophylly. Ecology 47: 992–1007.

Berg, R.Y. 1975. Myrmecochorous plants in Australia and their dispersal by ants. Aust. J. Bot. 23: 475–508.

Bond, P. & Goldblatt, P. 1984. Plants of the Cape Flora. A Descriptive Catalogue. J. S. Afr. Bot. Supplementary 13: 1–455.

Bond, W.J. & Breytenbach, G.J. 1985. Ants, rodents, and seed predation in Proteaceae. S. Afr. J. Zoo. 20: 150–154.

Bond, W.J. & Slingsby, P. 1983. Seed dispersal by ants in shrublands of the Cape Province and its evolutionary implications. S. Afr. J. Sci. 79: 231–233.

Bond, W.J., Yeaton, R. & Stock, W.D. 1991. Myrmecochory in Cape fynbos. In: Huxley, C.R. & Cutler, D.F. (eds.). Ant-Plant Interactions. Oxford University Press, pp. 448–462.

Bond, W.J. & Slingsby, P. 1983. Seed dispersal by ants in shrublands of the Cape Province and its evolutionary implications. S. Afr. J. Sci. 79: 231–233.

Boucher, C. & Moll, E.J. 1981. South African Mediterranean Shrublands. In: di Castri, F., Goodall, D.W. & Specht, R.L. (eds.). Ecosystems of the World 11. Mediterranean-type Shrublands. Elsevier, Amsterdam, pp. 233–248.

Breytenbach, G.J. 1988. Why are myrmecochorous plants limited to fynbos (macchia) vegetation types? S. Afr. For. J. 144: 3–5.

Campbell, B.M. 1985. A classification of the mountain vegetation of the fynbos biome. Mem. Bot. Surv. S. Afr. 50: 1–121.

Cowling, R.M. 1984. A syntaxonomic and synecological study in the Humansdorp region of the fynbos biome. Bothalia 15: 175–227.

Cowling, R.M. 1987. Fire and its role in coexistence and speciation in Gondwanan shrublands. S. Afr. J. Sci. 83: 106–112.

Cowling, R.M. & Bond, W.J. 1991. How small can reserves be? An empirical approach in Cape Fynbos. Biol. Cons. 58: 243–256.

Cowling, R.M. & Campbell, B.M. 1980. Convergence in vegetation structure in the Mediterranean communities of California, Chile and South Africa. Vegetatio 43: 191–197.

Cowling, R.M. & Holmes, P.M. 1992. Endemism and speciation in a lowland flora from the Cape Floristic Region. Biol J. Linn. Soc. 47: 367–383.

Cowling, R.M., Pierce, S.M. & Moll, E.J. 1986. Conservation and utilization of South Coast Renosterveld, an endangered South African vegetation type. Biol. Cons. 37: 363–377.

Cowling, R.M., Straker, C.J. & Deignan, M.T. 1990. Does microsymbiont-host specificity determine plant species turnover and speciation in Gondwanan shrublands? A hypothesis. S. Afr. J. Sci. 86: 118–120.

Davidson, D.W. & Morton, S.R. 1981. Myrmecochory in some plants (F. Chenopodiaceae) of the Australian arid zone. Oecologia 50: 357–366.

Fowler, C.W. & MacMahon, J.A. 1982. Selective extinction and speciation: their influence on the structure and functioning of communities and ecosystems. Am. Nat. 119: 480–498.

Givnish, T.J. 1987. Comparative studies of leaf form: assessing the relative roles of selective pressures and phylogenetic constraints. New Phytol. 106: 31–60.

Grubb, P.J. & Hopkins, A.J.M. 1986. Resilience at the level of the plant community. In: Dell, B., Hopkins, A.J.M. & Lamont, B.B. (eds.). Resilience in Mediterranean-Type Ecosystems. W. Junk, The Hague, pp. 21–38.

Hall, A.V. & Veldhuis, H.A. 1985. South African Red Data Book: Plants – fynbos and karoo biomes. S. Afr. Nat. Sci. Prog. Rep. 117. CSIR, Pretoria.

Harper, J.L. 1977. Population Biology of Plants. Academic Press, London.

Hartmann, H.E.K. 1991. Mesembryanthema. Contrib. Bol. Herb. 13: 75–157.

Hoffman, M.T. & Cowling, R.M. 1987. Plant physiognomy, phenology and demography. In: Cowling, R.M. & Roux, P.W. (eds.). The Karoo Biome: a preliminary synthesis. Part 2. Vegetation and History. S. A. Nat. Sci. Prog. Rep. 142. CSIR, Pretoria, pp. 1–34.

James, S. 1984. Lignotubers and burls – their structure, function and ecological significance in Mediterranean ecosystems. Bot. Rev. 50: 225–266.

Kilian, D. 1991. Seed and seedling ecology of two co-occurring ericoid fynbos shrub species. M.Sc. Thesis. University of Cape Town.

Knight, R.S. 1988. Aspects of plant dispersal in the southwestern Cape with particular reference to the roles of birds as dispersal agents. Ph.D. Thesis, University of Cape Town.

Kruger, F.J. 1979. South African Heathlands. In: Specht, R. L. (ed.). Ecosystems of the World 9A. Heathlands and Shrublands Descriptive Studies. Elsevier, Amsterdam, pp. 19–80.

Le Maitre, D. & Midgley, J.J. 1992. Reproductive ecology of fynbos plants. In: Cowling, R.M. (ed.). The Ecology of Fynbos – Nutrients, Fire and Diversity. Oxford University Press, Cape Town, pp. 135–174.

Levyns, M.R. 1929. Veld-burning experiments at Ida's Valley, Stellenbosch. Trans. Roy. Soc. S. Afr. 17: 61–92.

Levyns, M.R. 1935. Veld burning experiments at Oakdale, Riversdale. Trans. Roy. Soc. S. Afr. 23: 231–243.

Lewis, H. 1962. Catastrophic selection as a factor in speciation. Evolution 16: 257–271.

Linder, H.P. 1985. Gene flow, speciation and species diversity patterns in a species-rich area: the Cape Flora. In: Vrba, E.S. (ed.). Species and Speciation. Transvaal Museum, Pretoria, pp. 53–57.

Linder, H.P. & Ellis, R.P. 1990. Vegetative morphology and

inter-fire survival strategies in the Cape fynbos grasses. Bothalia 20: 91–103.

Linder, H.P. & Vlok, J. 1991. The morphology, taxonomy and evolution of *Rhodocoma* (Restionaceae). Plant Syst. Evol. 175: 139–160.

Major, J. 1988. Endemism: a botanical perspective. In: Myers, A.A. & Giller, P.S. (eds.). Analytical Biogeography. An Integrated Study of Animal and Plant Distributions. Chapman and Hall, New York, pp. 117–146.

Manders, P.T. 1990. Fire and other variables as determinants of forest/fynbos boundaries in the Cape Province. J. Veg. Sci. 1: 483–490.

Manders, P.T. & Richardson, D.M. 1992. Colonization of Cape fynbos communities by forest communities. For. Ecol. & Mgmt. 48: 277–293.

Midgley, J.J. 1987. Aspects of the evolutionary biology of the Proteaceae, with emphasis on the genus *Leucadendron* and its phylogeny. Ph.D. Thesis, University of Cape Town.

Midgley, J.J. & Bond, W.J. 1989. Leaf size and inflorescence size may be allometrically related traits. Oecologia 78: 427–429.

Milewski, A.V. & Bond, W.J. 1982. Convergence of myrmecochory in Mediterranean Australia and South Africa. In: Buckley, R.C. (ed.). Ant-plant interactions in Australia. Dr W. Junk, The Hague, pp. 89–98.

Mossop, M.K. 1989. Comparison of seed removal by ants in vegetation on fertile and infertile soils. Aust. J. Ecol. 14: 367–373.

Musil, C.F. 1991. Seed bank dynamics in sand plain lowland fynbos. S. Afr. J. Bot. 57: 131–142.

Phillips, J.F.V. 1931. Forest succession and ecology in the Knysna region. Bot. Surv. Mem. 14: 1–327.

Pierce, S.M. 1990. Pattern and process in South Coast Dune Fynbos: Population, community and landscape level studies. Ph.D. Thesis, University of Cape Town.

Pierce, S.M. & Cowling, R.M. 1991. Dynamics of soil-stored seed banks of six shrubs in fire-prone dune fynbos. J. Ecol. 79: 731–747.

Primack, R.B. 1987. Relationships among flowers, fruits and seeds. Ann. Rev. Ecol. Syst. 18: 409–430.

Raven, P.H. 1964. Catastrophic selection and edaphic endemism. Evolution 18: 336–338.

Rice, B.L.& Westoby, M. 1981. Myrmecochory in sclerophyll vegetation of the West Head, New South Wales. Aust. J. Ecol. 6: 291–298.

Slingsby, P. & Bond, W.J. 1981. Ants – friends of the fynbos. Veld & Flora 67: 39–45.

Slingsby, P. & Bond, W.J. 1985. The influence of ants on the dispersal distance and seedling recruitment of *Leucospermum conocarpodendron* (L.) Buek (Proteaceae). S. Afr. J. Bolt. 5: 30–34.

Stebbins, G.L. & Major, J. 1965. Endemism and speciation in the Californian flora. Ecol. Monogr. 35: 1–31.

Stock, W.D. & Bond, W.J. 1992. On the costs of ant, bird and wind-dispersal in fynbos Asteraceae. Presented paper of the 18th Annual Conference of the South African Association of Botanists, Durban.

Stock, W.D., Pate, J.S. & Delfs, J. 1990. Influence of seed size and quality on seedling development under low nutrient conditions in five Australian and South African members of the Proteaceae. J. Ecol. 78: 1005–1020.

Thompson, K. & Rabinowitz, D. 1989. Do big plants have big seeds? Am. Nat. 133: 722–728.

Thwaites, R.N. & Cowling, R.M. 1988. Soil-vegetation relationships on the Agulhas Plain, South Africa. Catena 15: 333–345.

Van Wilgen, B.W., Higgins, K.B. & Bellstedt, D.U. (1990). The role of vegetation structure and fuel chemistry in excluding fire from forest patches in the fire-prone fynbos shrublands of South Africa. J. Ecol. 78: 210–222.

Vrba, E.S. 1980. Evolution, species and fossils: how does life evolve? S. Afr. J. Sci. 76: 61–84.

Westoby, M., Hughes, L. & Rice, B.L. 1991. Seed dispersal by ants; comparing infertile with fertile soils. In: Huxley, C.R. & Cutler, D.F. (eds.). Ant-Plant Interactions. Oxford University Press, Oxford, pp. 434–447.

Westoby, M., Rice, B. & Howell, J. 1990. Seed size and plant growth form as factors in dispersal spectra. Ecology 71: 1307–1315.

Williams, I. 1982. Studies in the genera of the Diosmeae (Rutaceae): 14. A review of the genus *Diosma* L. J. S. Afr. Bot. 48: 329–407.

Willson, M.F., Rice, W.L. & Westoby, M. 1990. Seed dispersal spectra: a comparison of temperate plant communities. J. Veg. Sci. 1: 547–562.

Yeaton, R.I. & Bond, W.J. 1991. Competition between two shrub species: dispersal differences and fire promote coexistence. Am. Nat. 138: 328–341.

Index of key words

Page numbers are the first page of a paper in which the entry is discussed

Acacia scrub 47
Acridid grasshopper 105
Afromontane forest 151
Ants 73, 159
Aristotle 3

Biological control 83
Biology 3
Birds 47
Blue Tits 25
Botany 3

Cape Floral Kingdom 151
Carnivorous plants 83
Caterpillars 25
Community diversity 15
Community structure 37
Convergence 115
Convergent evolution 137
Core plants 125
Creosote bush 105

Dispersal 159
Disturbance 73
Diversification 159
Diversity 47

Environment-mediated interactions 73
Evergreenness 25

Flower colour 125
Fynbos 137

Generalists 63
Gophers 73
Grazing 93

Habitat mosaics 25
Heathlands 47, 151
Herbivory 83, 105, 115

Larrea tridentata 105
Leaf-form spectra 63

Mallee 47
Maquis 125
Mediterranean 125
Mediterranean Basin rangelands 93
Mediterranean ecosystems 115
Mediterranean shrublands 151
Mediterranean-type ecosystems 37
Meneris tulbaghia 137
Metameric architecture 115
Mulga 47
Mutualism 63, 83
Mycorrhizas 83

NDGA 105
Nectar 137

Plant-animal interactions 3, 63
Pollination 137
Pollination biology 125
Pollination market 125
Primary productivity 15
Psychophily 137

Regional peculiarities 125

Seasonal features 125
Serpentine grassland 73
Shrublands 47, 63
Soil fauna 37
Solar radiation 15
Source/sink 25
Species diversity 15
Species richness 37
Summergreenness 25
Sunbirds 137

Theophrastus 3
Trophic relations 83
Turnover 47

Vascular plants 15
Vegetation 93
Vertebrates 15

Zoology 3

Author index

Abbott, I. 41, 42, 43
Abrahamson, W.G. 112, 119
Acocks, J.P.H. 151, 152
Adams, R.J. 69
Aljaro, M.E. 116, 118, 119
Andersen, A.N. 160
Anderson, S.S. 112
Ankey, C.D. 31
Arcese, P. 31
Aristotle, 3, 4, 5, 6, 7, 8, 9, 10
Armstrong, J.A. 68
Arranz, R. 118
Attenborough, D. 93, 94, 100
Austin, M.P. 79
Avila, G. 116, 119
Axelrod, D.I. 16

Baker, H.G. 128, 143
Baker, I. 143
Bamberg, S.A. 105
Bandoli, J.H. 75
Barbour, M.G. 63, 105
Baumann, H. 8
Bawa, S.K. 128
Beadle, N.C.W. 162
Bean, A. 164
Beattie, A. 78, 85
Bell, A.D. 115
Bell, D.T. 40, 41, 43, 63, 64, 65, 67, 69
Bellairs, S.M. 67
Bellstedt, D.U. 166
Berg, R.Y. 79, 159
Bergstrom, G. 132
Berliner, R. 97
Bernard, G.D. 138
Bernays, E.A. 107, 108, 111, 113
Bernhardt, P. 10, 132, 137
Betts, M.M. 30, 31
Bigot, L. 40
Blandin, P. 39, 40, 41
Blondel, J. 25, 26, 27, 28, 29, 31, 32, 33, 34
Bodot, P. 40
Bohnstedt, C.F. 109
Bolton, M.P 21
Bond, P. 151, 159, 165
Bond, W.J. 78, 79, 137, 159, 160, 161, 162, 164, 165
Bonnier, J. 100
Bossi, L. 152
Boucher, C. 153, 154, 159
Bougher, N.L. 88
Box, T.W. 95, 96
Boyden, T.C. 146
Breytenbach, G.J. 159, 160, 162
Bristow, A. 8
Bryant, J.P. 112
Buckley, R.C. 67, 78, 79

Calder, M. 10
Calder, W.A. 31
Caldwell, M.M. 97
Campbell, B.M. 151, 152, 159, 160, 162, 166
Cansela de Fonseca, G.P. 37, 38, 42
Cassis, G. 88
Catling, P.C. 21, 22, 23
Chapin, F.S. 112
Chapman, R.F. 107, 108, 111, 113
Charnov, E.L. 32
Chessel, D. 34
Chew, A.E. 118
Chew, R.M. 118
Chiarello, N. 75, 78
China, W.E. 87
Christensen, N.L. 118
Christensen, P. 67, 85
Christensen, P.E. 84, 85
Claassens, A.J.M. 147
Clamens, A. 26
Clarke, S.A. 86
Cocks, M. 159, 163
Cody, M.L. 34, 47, 49, 50, 51
Cohen, D. 127
Coley, P.D. 112

Contreras, L.C. 75, 77, 78
Cook, J.B. 78
Corbet, S.A. 126
Cottrell, C.B. 137
Cowie, R.J. 30, 31
Cowling, R.M. 152, 159, 160, 161, 162, 164, 165, 166
Cox, G.W. 75, 77
Cramm, P. 26
Crawley, M.J. 94, 96, 100
Crespo, D.G. 99, 100
Cruden, R.W. 143, 146
Cunningham, G. 105
Cushman, J.H. 85

Daan, S. 27, 31
Dafni, A. 95, 125, 127, 128, 130, 131, 132, 137
Dan, J. 94
Daniel, T.L. 147
Darwin, C. 95
Davidson, D.W. 160
Davies, K.C. 78
Davis, S.D. 118
Deacon, H.J. 152
Deignan, M.T. 166
De Laet, J.V. 33
De Lange, C. 155
Delfs, J.C. 69, 162, 163
De Ridder, N. 100
Dervieux, A. 26, 28, 29, 32
Detling, J.K. 118
Dettmann, M.E. 15, 16
Diamond, J.M. 34
Dias, P.C. 25, 26, 32
Di Castri, F. 37, 38, 40, 41, 43, 94
Dickson, C.G.C. 147
DiFeo, D.R. 109
Dole, J. 108
Donnelly, D. 39, 40, 41
Downum, K.R. 107, 108, 110
Drent, R.H. 27, 31
Dronamraju, K.R.F. 146, 147
Du Merle, P. 26, 27, 29
Dukas, R. 126, 131
Dwyer, P.D. 47
Dyer, M.I. 118

Eckardt, F.E. 26
Edmonds, S.J. 21, 40
Edney, E.B. 32
Ehrlich, P.R. 77, 105
Elgueta, M. 118, 119, 120
Ellis, J.E. 94
Ellis, R.P. 164
Enders, F. 105, 106, 107, 108, 112
Endler, J.A. 138, 144
Eriksson, T. 145, 147
Etchegaray, J. 100, 118, 119
Etienne, M. 100
Evenari, M. 5
Everitt, B.S. 117

Faegri, K. 133, 139
Feeny, P.P. 26
Fenner, M. 10
Flores, E. 119
Floret, Ch. 26, 116
Foran, B.D. 93
Ford, H.A. 68
Fowler, C.W. 166
Franco, M. 117
Frankie, G.W. 131
Frazer, J.M. 118
Fretwell, S.D. 33, 34
Frias, D. 119
Friedel, M.H. 93
Frochot, B. 34
Fuentes, E.R. 100, 118, 119
Fuggle, R.F. 151

Gack, C. 132
Galan, M.J. 26, 116
Galen, C. 68
Galil, J. 9, 78
Garrison, G.A. 99
Gashwiller, J.S. 118
Gaubert, H. 26
George, A.S. 47
Ghabbour, S.I. 37, 38, 40, 41, 42
Gibb, J.A. 30, 31
Gibson, A. 106, 107
Giliommee, J.H. 39, 40, 41
Gillet, H. 99
Gilmore, A. 50
Ginocchio, R. 115, 117, 119
Givnish, T.J. 162
Goeden, R.D. 79
Goldblatt, P. 151, 159, 165
Golley, F.B. 118
Gomez, M. 116, 119
Gonzalez-Coloma, A. 105, 108, 109, 110, 111
Grant, K.A. 137
Grebbenikov, V.S. 126
Greenbaum, S. 132, 137
Greenfield, M.D. 107, 108, 110, 111, 113
Greenslade, P. 37, 39, 41
Gross, G. 43
Grove, T.S. 88
Groves, R. 4, 48
Grubb, P.J. 164
Grundy, R.I. 20, 23, 96

Gulmon, S.L. 105
Gutierrez, J. 78
Gutman, M. 96, 99, 100

Haber, W.W. 131
Haig, D. 67
Haimov, Y. 99
Halford, D.A. 64, 65
Hall, A.V. 147, 164
Hallam, N.D. 86
Halle, F. 115
Hamburg, S.P. 75
Hamilton, A.G. 86
Harper, J.L. 96, 115, 118, 120, 164
Hartley, P.H.T. 31
Hartmann, H.E.K. 160
Heddle, E.M. 17, 63
Heim, G. 26
Hendley, Q.B. 152
Henkin, Z. 99, 100
Herbel, C.H. 95, 96, 97, 99, 100
Hermann-Parker, S.M. 143, 146
Herrera, C. 127
Herrera, J. 125, 126, 127, 128, 130, 131, 132
Hickman, J.C. 75
Higgins, K.B. 166
Hilbert, D.W. 118
Hilborn, R. 27, 31
Hill, M.O. 21
Hilton-Taylor, C. 152
Hingston, F.J. 84, 85
Hinsley, S.A. 30, 31
Hobbs, R.J. 73, 75, 76, 77, 78, 80
Hoffman, A.E. 116, 119
Hoffman, A.J. 116, 119
Hoffman, M.T. 160
Holechek, J.L. 95, 96, 97, 99, 100
Holmes, P.M. 152, 160, 162, 164, 165, 166
Holzer, Z. 99, 100
Hopkins, A.J.M. 47, 164
Hopkins, B. 15
Howard, J.J. 109
Howell, J. 159, 162, 163, 165
Huenneke, L.F. 75
Huntley, B.J. 152
Huntly, N. 73, 77
Hurd, P.D. 128
Hussein, A.K.M. 38, 42

Immelmann, K. 31
Ingram, G.J. 47
Inouye, R. 73, 77
Isenmann, P. 26
Ivri, Y. 132, 137

Jakupcak, J. 108
James, S. 166
Jarman, M.L. 152
Jarvis, J.V.M. 78, 153
Jarzen, D.M. 15
Joel, D. 87
Joern, A. 107, 113
Johnson, N.D. 105
Johnson, S.D. 137, 138
Jordan, M. 116, 118
Joubert, C. 152, 153

Kaplan, Y. 96, 99
Karamaouna, M. 38, 39, 40, 41
Karr, J.R. 34
Keast, A. 47
Kerley, G. I.H. 67
Khanna, P. 79
Kikkawa, J. 47
Kilian, D. 162, 164, 166
Kimber, P. 67
King, J.R. 31
Kingsolver, J.G. 147
Kiortsis, B. 4, 5
Klein, D.R. 112
Klomp, H. 31
Koch, L. E. 39, 40, 41
Kohen, J.L. 67
Koide, R.T. 75
Kolb, G. 146
Krebs, J.R. 30, 32
Kruger, F.J. 155, 161, 166
Kullenberg, B. 132
Kummerow, J. 116, 125
Kuppers, M. 115
Kyriacou, X. 116

La Berge, W.E. 133
Lack, D. 29, 31, 32
La Marche, V.C. 118
Lambrechts, J.J. 152
Lamont, B.B. 83, 84, 85, 86, 88
Landsberg, J. 79
Lange, O.L. 97
Langedyk, W. 42
Lawrence, W. 118
Lebreton, Ph. 26
Le Floch, L.E. 26, 116
Legakis, A. 37, 39, 40, 41
Le Houerou, H.N. 93, 99
Le Maitre, D. 159, 160, 161, 162, 164, 166
Lepart, J. 100
Leprince, F. 26, 116
Le Roux, A. 116

Levin, D.A. 69
Levyns, M.R. 166
Lewis, H. 166
Lieberman, A.S. 93, 96, 97
Liengme, C.A. 154
Lifjeld, J.T. 31
Lightfoot, D.C. 106, 112
Lincoln, D.E. 105
Linder, H.P. 161, 164, 166
Lions, J.C. 43
Litav, M. 97
Loneragan, W.A. 64, 65, 69
Losito, L. 132, 137
Louda, S.M. 118
Lovegrove, B.G. 153
Lovegrove, B.P. 78
Lowell, P.H. 115, 117
Lowell, P.J. 115, 117
Loyd, J.W. 154
Lucas, H.L. 34
Lundberg, H. 145, 147
Lyons, M.N. 69

Mabry, T.J. 108, 109
MacArthur, R.H. 34, 69
MacInnes, C.D. 31
MacLead, N.D. 93
Magioris, S.N. 38, 39, 41, 42
Maistre, M. 26, 28, 29, 32, 33
Majer, J.D. 37, 39, 40, 41, 42, 43
Major, J. 164, 166
Malajczuk, N. 84, 85, 88
Malanson, G.P. 100
Manders, P.T. 160, 166
Manville, G.C. 69
Marchant, N.G. 47
Marcuzzi, G. 43
Marloth, 137
Marmari, A. 38, 43
Martin, T.E. 27, 31, 32
Martinez, E. 118, 119, 120
Martinez, J. 118
Mason, H. 153
Matthei, O. 79
Mazet, R. 26, 27
McAndrews, J.H. 69
McCleery, R.H. 31
McCrea, E.D. 112
Mc Donald, D.J. 155
McKenzie, B. 151
McMahon, J.A. 166
McNab, B.K. 77
McNaughton, S.J. 118
Meeuse, B. 8
Meinzer, F.C. 109, 110
Meister, H.P. 97
Menzel, R. 131
Mertens, 30, 31, 32
Methy, M. 26
Michael, P.W. 79
Michener, C.D. 126
Midgley, J.J. 159, 160, 161, 162, 164, 166
Mielke, H.W. 73
Mikhail, W.Z.A. 37, 38, 41
Milewski, A.V. 75, 77, 78, 79, 159, 160
Miller, M.A. 75
Mills, J.N. 118
Minnich, R.A. 63
Minot, E.O. 30, 31
Mispagel, M.E. 106, 112
Molina, J.D. 118
Moll, E.J. 151, 152, 153, 159, 164
Montenegro, G. 115, 116, 117, 118, 119, 120
Mooney, H.A. 73, 75, 76, 77, 80, 94, 105
Mordoundt, J.A. 65
Morgan, D.G. 17, 19
Morin, Y. 97
Morris, S. 8
Morrow, P.A. 105, 118
Morse, J. 79
Morton, A.G. 3, 4, 5, 10
Morton, S.R. 160
Mossop, M.K. 160
Mujica, A.M. 116, 119
Muller, C.H. 118
Musil, C.F. 164
Mustafa, J. 118

Naiman, R.J. 73
Naveh, Z. 93, 94, 95, 96, 97, 100
Nevo, E. 74, 77
Newport, M.E. 68
Nilsen, E.T. 109, 110
Noitsakis, B. 99
Noy-Meir, I. 96, 97, 99, 100
Nur, N. 31

Oechel, W.C. 105, 118
Oldeman, A.A. 115
Oldeman, R.A.A. 115
Oliveto, E.P. 108
Opler, P.A. 131
Orians, G.H. 118
Orshan, G. 97, 116
Osbourne, J.L. 126
Osorio, R. 116, 119
O'Toole, C. 125, 126, 127, 130, 132, 133, 137
Otte, D. 105, 106, 107, 108, 112, 113

Owen, D.F. 29, 118

Pantis, J.D. 37, 39, 41
Papageorgiou, N. 96
Papanastasis, V.P. 93, 99, 100
Paraschi, L. 39, 40, 41
Parkes, D.M. 86
Partridge, L. 30
Pashby, A.S. 42
Pate, J.S. 69, 162, 163
Paulus, H.H. 132
Percival, M.S. 146
Perevolotsky, A. 94, 99, 100
Perret, Ph. 26, 28, 29, 32, 33
Perrins, C.M. 26, 31
Perry, P. 116, 154
Petanidou, Th. 126, 127, 128
Pettifort, R.A. 31
Phillips, J.F.V. 160
Pianka, E.A. 69
Pieper, R.D. 95, 96, 97, 99, 100
Pierce, S.M. 159, 160, 162, 164, 166
Pignatti, S. 125
Pimm, S.L. 79
Poaini, A. 118
Poissonet, P. 100
Pons, A. 16
Postle, A.C. 38, 40, 41, 43
Prevost, M.F. 115
Primack, R.B. 162
Pulliam, H.R. 32
Putwain, P.D. 120

Quezel, P. 125, 133
Quinn, R. 118

Rabenbold, K.N. 29, 30, 31
Rabinovich-Vin, A. 97
Rabinowitz, D. 164
Radea, K. 38, 39, 43
Ralph, C.S. 84, 85
Raven, P.H. 137, 147, 166
Raw, A. 126
Rebelo, A.G. 152
Recher, R. 50
Rehn, J.A.G. 107
Reig, O.A. 78
Rhoades, D.F. 108, 109
Rice, B. 67
Rice, W.L. 159, 160, 161, 162, 163, 165
Richardson, D.M. 160, 166
Ridley, H.N. 153
Rizk, M.A. 37, 41
Rodriguez, E. 108
Romane, F. 26, 100, 116
Roubik, D.W. 126
Royama, T. 29, 30, 31
Rozenzweig, D. 97
Rundel, P.W. 105, 106, 107, 108, 109, 110
Russell, M.C. 87, 88
Rutherford, M.C. 151

Safriel, U. 96
Saiz, F. 39, 43
Sakakibara, M. 109
Sakellariou, B. 93, 97
Sanford, S. 94
Sattle(r), P.S. 47
Saunders, D.A. 83, 84
Sauvezon, R. 26
Scharloo, W. 32
Scherer, C. 146
Schowalter, T.D. 107
Schultz, J.C. 105, 106, 107, 108, 112
Schulze, B.R. 151
Scott, J.K. 84
Seginer, I. 97
Seigler, D.S. 108, 109
Sela, Y. 95
Seligman, N.G. 93, 99, 100
Selten, R. 127
Serey, I. 118
Sgardelis, J. 37, 39, 41, 42
Shachori, A. 97
Shakir, S.H. 37, 38
Sharifi, M.R. 105, 109, 110
Shelley, J.M. 67
Shelly, T.E. 107, 108, 110, 111, 113
Shepherd, H.R. 99
Shmida, A. 96, 125, 126, 127, 130, 131, 132, 137
Shortridge, G.C. 78
Simonetti, J.A. 118
Simpson, S.J. 107
Singer, M.C. 77
Slagsvold, T. 31
Slingsby, P. 159, 160, 161, 164, 165
Smith, A.D. 95, 96
Smith, J.N.M. 31
Smitheman, J. 154
Snow, D.W. 30
Solbrig, O.T. 118
Solervicens, J. 119
Specht, A. 15, 16, 17, 20, 23, 96
Specht, M.M. 40
Specht, R.L. 15, 16, 17, 19, 20, 21, 22, 23, 47, 48, 96
Springett, J.A. 41
Stafford-Smith, D.M. 93
Stamou, G.P. 37, 39, 41

Stebbins, G.L. 132, 166
Stewart, L. 85
Stiles, F.G. 68
Stock, W.D. 159, 160, 161, 162, 163, 164, 165
Stoddard, L.A. 95, 96
Straker, R.M. 166
Strauss, S.Y. 83
Strijkstra, A. 26, 31
Swift, D.M. 94, 118
Swihart, C.A. 146
Swihart, S.L. 146

Talamucci, P. 100
Taylor, J. 87
Tenhunen, J.D. 97
Thanos, C.A. 3, 5
Theophrastus 3, 4, 5, 6, 7, 8, 9, 10
Thiault, M. 100
Thirgood, J.V. 93, 96
Thompson, D'Arcy W. 4
Thompson, K. 164
Thwaites, R.N. 166
Tinbergen, L. 26, 30
Tomaselli, R. 93
Tomlinson, P.B. 115
Torok, J. 30, 31
Trabaud, L. 100
Trihas, A. 41
Trimen, R. 137
Tsiouvaras, C.N. 99

Van Balen, J.H. 26, 30, 31, 32
Van der Moezel, P.G. 69
Van der Pijl, L. 133, 139, 153
Van Leeuwen, S.J. 83, 84
Van Noordwijk, A.J. 32
Van Wilgen, B.W. 166
Veldhuis, H.A. 147, 164
Vitali-Di Castri, V. 37, 38, 40, 41, 43
Vlok, J. 166
Vogel, S. 132, 146
Vokou, D. 126, 128
Vrba, E.S. 161, 166

Wagenaar, K.T. 100
Walker, B.H. 100
Walkowiak, A.M. 118
Walter, H. 95
Waser, N.M. 68
Watkinson, A.R. 115, 118
Watt, A.S. 95
Weevers, T. 131
Weigert, R.G. 118
Weis, A.E. 119
Werger, M.J.A. 151
Werker, E. 127
Westerkamp, C. 132
Westfall, R.H. 151
Westman, B. 34
Westman, W.E. 15, 23, 100
Westoby, M. 67, 159, 160, 161, 162, 163, 165
Whelan, R.J. 42
White, J. 115, 118
Whitford, W.G. 106, 107, 112
Whittaker, R.H. 96
Wiklund, C. 145, 147
Williams, G.C. 32
Williams, I.H. 126, 164
Williams, K.S. 105
Williams, P. 126
Williamson, M. 34
Wills, R.T. 69
Willson, M.F. 159, 160, 161
Wilson, A.D. 93
Wisdom, C.S. 108, 109
Wyatt, T. 107, 108, 111, 113

Yeates, D. 86
Yeaton, R.I. 159, 160, 161, 164, 165
Yom-Tov, Y. 27, 31
Yonatan, R. 99

Zandt, H. 26, 31
Zimmer, K.A. 68
Zohar, Y. 97

Systematic index

Abies 10
Acacia 47, 48, 49, 60, 67
Acacia aneura 47, 48, 49
Acacia baileyana 68
Acacia celastrifolia 65
Acacia cyclops 67, 68
Acacia decurrens 68
Acacia extensa 65, 68
Acacia glauceptera 68
Acacia greggii 109
Acacia horridula 68
Acacia lasiocalyx 68
Acacia lasiocarpa 68
Acacia melanoxylon 67, 68
Acacia pendula 67, 68
Acacia podalyrifolia 67, 68
Acacia pulchella 68
Acacia redolens 68
Acacia rostellifera 65, 66
Acacia steedmanii 68
Acacia vestita 68
Acanthagenys rufogularis 55
Acanthiza apicalis 55
Acanthiza chrysorrhoa 55
Acanthiza inornata 50, 55
Acanthiza pusilla 56
Acanthiza robustirostris 55
Acanthiza uropygialis 55
Acanthorhynchus superciliosus 55
Acarina 38, 39, 86, 106
Acer monspessulanum 25
Acer obtusifolium 129
Acer opalus 25
Aconitum 7
Adenanthos cygnorum 65
Adenostoma 18
Adonis allepica 132
Agaontidae 8
Agathosma 164, 165
Agathosma apiculata 164
Agathosma stenopetala 164
Aizoaceae 153
Allium neapolitanicum 128
Allium subhirsutum 128
Allocasuarina 65, 66
Amaryllidaceae 139, 145, 146
Amphicoma 132, 137
Amphipoda 39
Amygdalus communis 128
Anabasis 37, 38
Anacardiaceae 160
Andrena 126
Andrena carmela 132, 133
Andrenidae 126
Anellida 37, 38, 39
Anemone coronaria 132
Anthocercis littorea 65
Anthochaera carunculata 55
Anthochaera chrysoptera 50, 54
Anthocopa 126
Anthophora 131, 133
Anthophoridae 126
Anthospermum aethiopicum 154
Anthus novaeseelandiae 56
Aphelocephala leucopsis 55
Apis melifera 69, 126
Arachnida 37
Araneae 38, 39
Arbutus andrachne 129
Arbutus unedo 38
Aristolochia billardieri 130
Aristolochia sempervirens 130
Artamus cinereus 55
Artamus personatus 55
Artemisia tridentata 99
Arthrophora 83, 84
Aspalathus 153, 154, 165, 166
Asparagus aphyllus 130
Asparagus asparagoides 65
Asphodelaceae 146
Asphodelus 9, 133
Asphodelus aestivus 131, 132

Asphodelus fistulosus 65, 66, 68
Asteraceae 126, 153, 159, 160, 161, 165
Astroloma ciliatum 65
Astroma 108
Astroma quadrilobatum 107
Atriplex 107

Baccharis linearis 119
Baccharis pilularis ssp. *consanguinea* 80
Badisis 88
Badisis ambulans 86
Baeckaea camphorasmae 65
Bahia ambrosioides 119
Banksia 48, 84
Banksia tricuspis 83, 84, 88
Barnardius zonarius 55
Bathyergus 77
Bathyergus suillus 77
Bauhinia purpurea 68
Bettongia 85
Bettongia penicillata 84, 85
Blastophaga psenes 9
Bombyliidae 127, 128, 130
Bootettix 106, 107
Bootettix argentatus 106, 107, 108, 110, 112, 113
Boraginaceae 131
Bossiaea eriocarpa 65, 66
Bossiaea linophylla 65
Bossiaea ornata 65, 68
Brodiea 76
Bromus 76
Bromus mollis 75, 77
Brunsvigia marginata 139, 141, 143, 144, 145
Brunsvigia orientalis 144, 146
Bryonia cretica 130
Bryonia syriaca 130
Burchardia multiflora 65
Burchardia umbellata 68
Byblis 87, 88
Byblis gigantea 87

Cacatua roseicapilla 55
Cacatua sanguinia 50
Caesalpinia pulcherrima 137
Calamanthus fuliginosus 50
Calandrina 78
Calicotome 133
Callitris preissii 65
Calothamnus sanguineus 65
Calycadenia 76
Calycadenia multiglandulosa 77
Calyptorhynchus 84
Calyptorhynchus baudinii 50
Calyptorhynchus funereus 84
Calyptorhynchus funereus latirostris 83
Cannomois 162
Capra aegagrus cretica 8, 96
Carabidae 41
Carthamus 9
Ceanothus 18
Cedrus atlantica 25
Celastraceae 160
Celtis australis 129
Cephalanthera 132
Cephalotus 88
Cephalotus follicularis 86
Ceratocaryum 162
Ceratonia siliqua 128, 129
Ceratopogonidae 86
Cercidium floridum 109
Cercis siliquastrum 128, 129
Certhionyx niger 55
Certhionyx variegatus 55
Chacophaps indica 54
Chalcidoidae 118
Chalicodoma 126, 131, 133
Chasmanthe aethiopica 144, 146
Chasmanthe floribunda 140, 146
Chilopoda 38, 39, 40
Chilopsis linearis 109
Chrysanthemoides incana 154
Chrysomelidae 119
Cibolacris parviceps 108
Cicadellidae 106
Cicer arietinum 6
Cinclorhamphus mathewsi 56
Cinclosoma cinnamomeum 55
Cisticola exilis 54
Cistus 125, 133
Citrullus lanatus 68
Clematis cirrhosa 130
Clematis flammula 130
Clematis pubescens 65
Cliffortia 164, 165
Colchicum 127
Coleoptera 28, 38, 39, 40, 43, 86, 106, 119, 128, 130
Collembola 38, 39, 43
Colliguaja odorifera 115, 118, 119, 120
Colluricincla harmonica 50, 55, 56
Colutea arborescens 9
Conomyrma chilensis 119
Conopophila albogularis 50, 54
Conostylis setigera 65, 66
Copepoda 86
Coracina novaehollandiae 56
Coracina tenuirostris 56
Coridothymus capitatus 9, 131
Corvus bennetti 55
Corvus coronoides 50, 56
Corvus orru 50

Cracticus torquatus 56
Crassula coccinea 138, 139, 140, 143, 144, 145
Crassulaceae 139, 145
Crataegus 128
Crataegus aronia 129
Crataegus azarolus 129
Crocus 127
Cruciferae 126
Crustacea 37
Cryptocarya alba 119
Cryptomys 77
Cryptomys damarensis 77
Cryptomys hottentotus 77, 78
Cubitalia 131
Cucumis melo var. *cantalupensis* 68
Cunonia capensis 155
Cyathochaeta avenacea 65
Cyclamen 133
Cyperaceae 162
Cyrtanthus elatus 139, 141, 145
Cyrtanthus guthrei 139, 141, 143, 144, 145, 146
Cyrtanthus montanus 139
Cyrtanthus ventricosus 146
Cyrtopeltis 88
Cyrtopeltis droserae 87
Cyrtopeltis russellii 87
Cytisus 125

Dactylorrhiza 132
Dampiera lavandulacea 65
Dasyphelea 86
Daviesia decurrens 65
Daviesia juncea 65
Dendromecon rigida 79
Dermaptera 28, 38, 39
Dianella revoluta 65
Dictyoptera 38, 39
Dicyphini 87
Diosma 164
Diplopoda 37, 38, 39, 40
Diplura 38, 39
Diptera 28, 38, 39, 43, 86, 119, 128
Disa ferruginea 139, 143, 144, 145, 146
Disa porrecta 139
Disa uniflora 137, 139, 142, 143, 144, 145
Dromaius novaehollandiae 56
Drosera 87, 88
Dryandra carduacea 65
Drymodes brunneopygia 56

Ebenaceae 160
Elaphoidella 86
Elytropappus 154, 164
Elytropappus rhinocerotis 152, 153
Embioptera 38, 39
Eopsaltria australis 56
Ephedra foemina 130
Ephthianura tricolor 55
Eremophila glabra 65
Erica 146, 165, 166
Erica abietina 144
Erica hispidula 155
Ericaceae 146, 153, 164, 165
Eriocephalus africanus 153
Eriolobus trilobatus 128, 129
Eucalyptus 22, 38, 47, 48, 85
Eucalyptus calophylla 84, 85
Eucalyptus erythrocorys 68
Eucalyptus marginata 38, 39, 68
Eucalyptus wandoo 68
Eucera 126, 131, 133
Euclea racemosa 154
Eulophidae 118
Euonymus europaeus 7
Euphorbiaceae 118
Euphydras 77
Euphydras editha 77
Euryops 165
Exurus colliguayae 118, 119, 120

Fabaceae 84, 165
Ferula tingitana 8
Ficinia 162
Ficus carica 8
Fuchsia lycioides 119

Galium samuelssonii 128
Garrulus glandarius 9
Gastrolobium 66, 84, 85
Gastrolobium bilobum 65, 66, 67, 68, 84, 88
Gastrolobium calycinum 65, 66
Gastrolobium trilobium 65, 66
Genista 133
Georychus campensis 78
Gerygone fusca 55
Gladiolus bonaespei 144
Gladiolus cardinalis 139, 142, 145
Gladiolus cruentus 139
Gladiolus nerinoides 139, 140, 143, 144, 146
Gladiolus priorii 144, 146
Gladiolus saundersoniae 139
Gladiolus segetum 9
Gladiolus sempervirens 139, 146
Gladiolus stefaniae 139, 145, 146
Gladiolus stokei 139
Gladiolus watsonius 146
Glaphyridae 132
Glischrocaryon aureum 65

Gnidia 165
Gompholobium preissii 65
Gompholobium tomentosum 67, 68
Grallina cyanoleuca 56
Gryllidae 119
Gymnorhina tibicen 55
Gyrostemon subnudus 65

Habropoda 131
Haemanthus coccinea 146
Haemanthus rotundifolius 144, 146
Hakea ambigua 65
Hakea lissocarpa 68
Hakea trifurcata 65
Hakea undulata 65
Halictidae 126
Halictus 126, 131
Harpabittacus australis 87
Hedera helix 9, 130
Helichrysum vestitum 155
Heliconisus charitonius 146
Helleborus cyclophyllus 7
Hemiptera 28, 38, 39, 40, 41
Heteroptera 106
Hibbertia cunninghamii 65
Hibbertia racemosa 65
Homoptera 106
Hoplitis 126, 133
Hymenoclea salsola 109
Hymenogaster 85
Hymenoptera 28, 38, 39, 40, 86, 118, 119
Hypocalymma angustifolium 65, 68
Hypocalymma robustum 68
Hypodiscus 162

Insecta 37
Inula graveolens 8
Inula viscosa 8
Iridaceae 139, 145, 146
Iridomyrmex conifer 86
Ischyrolepis capensis 153
Isoodon 85
Isopoda 38, 39, 40
Isoptera 38

Jacksonia restioides 65
Juniperus 41
Juniperus oxycedrus 129
Juniperus phoenicea 38, 39

Kageneckia oblonga 119
Kniphofia 146

Lalage sueurii 56
Lamiaceae 127, 131
Lantana camara 137, 146
Larrea 105, 106, 107, 108, 109, 110, 112, 113
Larrea cuneifolia 107
Larrea tridentata 105, 106, 107, 108, 109, 111, 112, 113
Lasioglossum 126
Lasiopetalum molle 65
Lasiopetalum oppositifolium 65
Lasthenia 76
Lasthenia californica 75, 77
Lathyrus ochrus 9
Laurus nobilis 7, 129
Lavandula 125, 133
Lavandula stoechas 131
Lepidoptera 28, 38, 39, 86, 106, 119
Lepidosperma scabrum 65
Leucadendron 165, 166
Leucadendron laureolum 154
Leucopogon capitalatus 65
Leucopogon pulchella 65
Leucopogon verticellatus 65
Leucoryne ixiodes 78
Leucospermum 153, 165
Lichenostomus leucotis 55
Lichenostomus virescens 50, 55
Lichmera indistincta 50, 55
Ligurotettix coquilletti 107, 108, 110, 111, 112, 113
Lilium candidum 95
Linaria 133
Lithrea caustica 119
Lomandra effusa 65
Lomandra hermaphrodita 65
Lomandra preissii 65
Lonicera etrusca 130
Loranthus europaeus 10
Loxocarya fasciculata 65
Loxocarya flexuosa 65
Lupinus albus 6
Lycium 107

Macrophalia 119
Macropus eugenii 65, 66
Macropus fulginosus 65, 66
Macrozamia reidleii 65
Malurus cyaneus 56
Malurus lamberti 50, 55, 56
Malurus leucopterus 55
Malurus melanocephalus 50, 56
Malurus splendens 55
Manorina flavigula 55
Mastersiella 162
Meberus tykvaggua 140
Mecoptera 39
Medicago arborea 7

Medicago sativa 7, 9
Megachile 126, 133
Megachilidae 126
Megalurus timoriensis 50, 54
Meliphaga gracilis 56
Meliphaga notata 54
Meliphagidae 50
Melithrepus brevirostris 55
Melittoides 132
Melittoides melittoides 133
Melopsittacus undulatus 55
Membracidae 106
Meneris 137, 138, 139, 140, 142, 143, 144, 145, 146, 147
Meneris tulbaghia 137, 138, 139, 141, 143, 145, 146
Mesembryanthemaceae 160
Mesomalaena tetragona 65
Mesophellia 84, 85
Metalasia 164
Metalasia muricata 164
Micranthus junceus 78
Microeca leucophaea 55
Micropezidae 86
Mimetes 165
Mirbelia ramulosa 65
Miridae 87, 106
Mollusca 37, 38, 39
Muehlenbeckia hastulata 119
Muraltia 164, 165
Muraltia squarrosa 164
Myriapoda 37
Myrtaceae 84
Myrtus communis 9, 129
Mythiocoymia 107

Narcissus 133
Nematoda 37, 38, 86
Nerine sarniensis 138, 139, 143, 144, 145
Nerium odorum 7
Nerium oleander 132
Neurachne alopecuroidea 65
Neuroptera 38
Nymphicus hollandicus 55

Ocyphaps lophotes 55
Olea europaea 128, 129
Olea europaea ssp. *oleaster* 7
Oleaceae 160
Oligochaeta 86
Opercularia vaginata 66
Operophtera brumata 29
Ophrys 132
Opiliones 38, 39
Orchidaceae 132, 139, 145, 153
Orchis 132
Orchis caspia 132
Oreoica gutteralis 55
Oribatidae 86
Origanum 8
Origanum dictamnus 7
Orthoptera 28, 38, 39, 106, 119
Osteospermum 160
Oxalis 78
Oxylobium 66

Pachycephala 56
Pachycephala griseiceps 54
Pachycephala pectoralis 56
Pachycephala rufiventris 55, 56
Pancratium parviflorum 127
Papaver 132, 133
Papaver rhoeas 9
Pardalotus striatus 55
Pardalotus xanthopygus 55
Paronomus 165
Parus c. *ogliastrae* 34
Parus caeruleus 25
Passerina 164
Passerina paleaceae 162, 164
Passerina vulgaris 164
Pauropoda 38
Persoonia longifolia 66
Petalonyx thurberi 109
Petroica goodenovii 55
Petroica multicolor 50
Petrophile serruriae 66
Peumus boldus 119
Phaps chalcoptera 56
Phaps elegans 56
Phillyrea latifolia 129
Phoenix canariensis 68
Phoenix dactylifera 8
Phylica 153, 164, 165
Phylica ericoides 162, 164
Phylica stipularis 164
Phylidonyris 50
Phylidonyris albifrons 50, 55
Phylidonyris melanops 50, 55
Phylidonyris nigra 54, 55
Phylidonyris novaehollandiae 50, 54, 55
Phylinae 87
Phyllanthus calycinus 66, 68
Pieridae 146
Pieris brassicae 146
Pinus 10
Pinus halepensis 38, 43
Pisolithus 85
Pistacia 128

Pistacia lentiscus 7, 129
Pistacia palaestina 129
Pistacia terebinthus 10
Pisum 133
Pittosporum phyllaeroides 68
Plantago 76, 77
Plantago erecta 75, 77, 78
Platycercus elegans 56
Poaceae 159
Poephila guttata 55
Polistes 133
Polygalaceae 165
Pomatostomus superciliosus 55
Prasium majus 130
Procalus lenzi 119
Procalus malaisei 119
Protea 164
Protea laurifolia 155
Proteaceae 48, 83, 153, 155, 161, 165, 166
Protura 38
Prunus amygdalus 9
Prunus ursina 128, 129
Pscocoptera 37, 38, 39, 40
Psephotus varius 55
Pseudoccoccidae 106
Pseudoscorpiones 38, 39
Psophodes cristatus 55
Psophodes olivaceus 54
Psorothamnus spinosus 109
Psyllidae 119
Putterlickia pyracantha 154
Pyrus amygdaliformis 9
Pyrus syriaca 128, 129

Quercus 7, 9, 10, 128
Quercus boissierii 129
Quercus calliprinos 94, 129
Quercus coccifera 7, 10, 18, 38, 39
Quercus ilex 18, 25, 26
Quercus ithaburensis 129
Quercus pubescens 25
Quillaja saponaria 119

Rachiptera limbata 119
Ramsayornis modestus 50, 54
Ranunculus asiaticus 132
Raphanus sativus 8
Rattus 85
Restionaceae 153, 155, 162
Rhamnaceae 165
Rhamnus alaternus 128, 129
Rhamnus lycoides ssp. *graeca* 129
Rhamnus punctata 129
Rhipidura fuliginosa 55
Rhipidura leucophrys 56
Rhus 152
Rhus coriaria 129
Rhus glauca 154
Rhus laevigata 154
Rhus lucida 154
Ricinus communis 68
Rosa canina 129
Rosaceae 165
Rubus canescens 130
Ruta graveolens 7
Rutaceae 165

Saccharum 9
Salvia 133
Salvia fruticosa 131, 132
Salvia viridis 7
Santalaceae 165
Satyridae 137
Scarabaeidae 41
Schinus polygamus 119
Schizostylis coccinea 139, 141, 144, 145
Schoenus cyperacea 66
Scleroderma 85
Scorodosma foetida 7
Scrophularia 132
Scrophularia rubricaulis 132, 133
Semiothisa 106
Semiothisa larreana 106
Sericornis brunneus 55
Sericornis frontalis 54, 55
Serruria 164, 165
Sesamum indicum 7
Setocaris bybliphilus 87
Setocoris 88
Shiphonaptera 39
Silene 133
Silene italica 128
Silene trinervis 128
Silene vulgaris 128
Silybum marianum 79
Sisymbrium polyceratium 7
Sitanion jubatum 75
Smicrornis brevirostris 55, 56
Smilax aspera 130
Solanum symonii 66
Sollya heterophylla 68
Solpugida 38, 39
Spalacopus cyanus 78
Spalax ehrenbergi 78
Spartium 133
Spartium junceum 9
Sphingidae 127, 128, 130
Staphylinidae 41

Sternbergia clusiana 127
Stipa flavescens 66
Stipa pulchra 75
Stipiturus malachurus 50, 54, 55
Stirlingia latifolia 68
Stoebe 164
Strepera versicolor 55
Strepsiptera 39
Stylidium affine 66
Stypandra imbricata 66
Styrax officinalis 8, 129
Symphyla 38, 39
Synhalonia 131
Syrphidae 127, 128, 130

Tainarys sordida 119
Talguenea quinquinervia 119
Talpa europaea 9
Tamus communis 130
Tamus orientalis 130
Taxus baccata 7
Testacea 8
Tetraria 162
Tetraria octandra 66
Tetratheca confertifolia 66
Teucrium 133
Teucrium polium 8
Thapsia garganica 7
Thesium 165
Thomasia cognata 66
Thomomys bottae 73
Thymelaea 37, 38, 41
Thymeleaceae 165
Thymus 9, 125, 133
Thysanoptera 37, 38, 39, 106
Thysanura 38, 39, 40
Tilia europaea 7
Torticidae 83
Torymus laetus 118, 119, 120
Trachyandra divericata 66
Trachymene octandra 68
Trevoa trinervis 119
Tribonanthes uniflora 66
Trichodere cockerelli 50, 54
Trichoglossus haematodus 56
Trichoptera 39
Trigonella graeca 9
Tritoniopsis 146
Tritoniopsis burchelli 146
Tritoniopsis leslei 139
Tritoniopsis longituba 139
Tritoniopsis nervosa 146
Tritoniopsis triticea 143, 144, 145, 146
Trymalium ledifolium 68
Trymalium spatulatum 68
Tulipa agenensis 132
Turnix maculosa 50, 54

Ulmus glabra 7
Ulmus minor ssp. *canescens* 129

Verbascum sinuatum 7
Veromescor andrei 78
Vespidae 128, 130
Vespula 133
Viburnum tinus 129
Vicia ervilia 6, 7, 8
Vicia faba 7, 9
Vicia sativa 7
Viscum album 10
Vitex agnus-castus 9
Vulpia 76
Vulpia microstachys 77

Watsonia 146
Watsonia tabularis 146
Willdenowia 162
Witsenia maura 146

Xanthorrhoea preissii 66

Zosterops lateralis 50, 55
Zosterops pallidus 51
Zygophyllaceae 160
Zygophyllum 160

Tasks for vegetation science

1. E.O. Box: *Macroclimate and plant forms*. An introduction to predictive modelling in phytogeography. 1981 ISBN 90-6193-941-0
2. D.N. Sen and K.S. Rajpurohit (eds.): *Contributions to the ecology of halophytes*. 1982 ISBN 90-6193-942-9
3. J. Ross: *The radiation regime and architecture of plant stands*. 1981 ISBN 90-6193-607-1
4. N.S. Margaris and H.A. Mooney (eds.): *Components of productivity of Mediterranean-climate regions*. Basic and applied aspects. 1981 ISBN 90-6193-944-5
5. M.J. Müller: *Selected climatic data for a global set of standard stations for vegetation science*. 1982 ISBN 90-6193-945-3
6. I. Roth: *Stratification of tropical forests as seen in leaf structure* [Part 1]. 1984 ISBN 90-6193-946-1 *For Part 2, see Volume 21*
7. L. Steubing and H.-J. Jäger (eds.): *Monitoring of air pollutants by plants*. Methods and problems. 1982 ISBN 90-6193-947-X
8. H.J. Teas (ed.): *Biology and ecology of mangroves*. 1983 ISBN 90-6193-948-8
9. H.J. Teas (ed.): *Physiology and management of mangroves*. 1984 ISBN 90-6193-949-6
10. E. Feoli, M. Lagonegro and L. Orláci: *Information analysis of vegetation data*. 1984 ISBN 90-6193-950-X
11. Z. Šesták (ed.): *Photosynthesis during leaf development*. 1985 ISBN 90-6193-951-8
12. E. Medina, H.A. Mooney and C. Vázquez-Yánes (eds.): *Physiological ecology of plants of the wet tropics*. 1984 ISBN 90-6193-952-6
13. N.S. Margaris, M. Arianoustou-Faraggitaki and W.C. Oechel (eds.): *Being alive on land*. 1984 ISBN 90-6193-953-4
14. D.O. Hall, N. Myers and N.S. Margaris (eds.): *Economics of ecosystem management*. 1985 ISBN 90-6193-505-9
15. A. Estrada and Th.H. Fleming (eds.): *Frugivores and seed dispersal*. 1986 ISBN 90-6193-543-1
16. B. Dell, A.J.M. Hopkins and B.B. Lamont (eds.): *Resilience in Mediterranean-type ecosystems*. 1986 ISBN 90-6193-579-2
17. I. Roth: *Stratification of a tropical forest as seen in dispersal types*. 1987 ISBN 90-6193-613-6
18. H.-G. Dässler and S. Börtitz (eds.): *Air pollution and its influence on vegetation*. Causes, Effects, Prophylaxis and Therapy. 1988 ISBN 90-6193-619-5
19. R.L. Specht (ed.): *Mediterranean-type ecosystems*. A data source book. 1988 ISBN 90-6193-652-7
20. L.F. Huenneke and H.A. Mooney (eds.): *Grassland structure and function*. California annual grassland. 1989 ISBN 90-6193-659-4
21. B. Rollet, Ch. Högermann and I. Roth: *Stratification of tropical forests as seen in leaf structure*, Part 2. 1990 ISBN 0-7923-0397-0
22. J. Rozema and J.A.C. Verkleij (eds.): *Ecological responses to environmental stresses*. 1991 ISBN 0-7923-0762-3
23. S.C. Pandeya and H. Lieth: *Ecology of Cenchrus grass complex*. Environmental conditions and population differences in Western India. 1993 ISBN 0-7923-0768-2
24. P.L. Nimis and T.J. Crovello (eds.): *Quantitative approaches to phytogeography*. 1991 ISBN 0-7923-0795-X
25. D.F. Whigham, R.E. Good and K. Kvet (eds.): *Wetland ecology and management*. Case studies. 1990 ISBN 0-7923-0893-X

Tasks for vegetation science

26. K. Falinska: *Plant demography in vegetation succession.* 1991 ISBN 0-7923-1060-8
27. H. Lieth and A.A. Al Masoom (eds.): *Towards the rational use of high salinity tolerant plants*, Vol. 1: Deliberations about high salinity tolerant plants and ecosystems. 1993 ISBN 0-7923-1865-X
28. H. Lieth and A.A. Al Masoom (eds.): *Towards the rational use of high salinity tolerant plants*, Vol. 2: Agriculture and forestry under marginal soil water conditions. 1993 ISBN 0-7923-1866-8
29. J.G. Boonman: *East Africa's grasses and fodders.* Their ecology and husbandry. 1993 ISBN 0-7923-1867-6
30. H. Lieth and M. Lohmann (eds.): *Restoration of tropical forest ecosystems.* 1993 ISBN 0-7923-1945-1
31. M. Arianoutsou and R.H. Groves (eds.): *Plant-animal interactions in Mediterranean-type ecosystems.* 1994 ISBN 0-7923-2470-6

KLUWER ACADEMIC PUBLISHERS – DORDRECHT / BOSTON / LONDON